Philosophical Engineering

♇METAPHILOSOPHY

METAPHILOSOPHY SERIES IN PHILOSOPHY

Series Editors: Armen T. Marsoobian and Eric Cavallero

The Philosophy of Interpretation, edited by Joseph Margolis and Tom Rockmore (2000)

Global Justice, edited by Thomas W. Pogge (2001)

Cyberphilosophy: The intersection of Computing and Philosophy, edited by James H. Moor and Terrell Ward Bynum (2002)

Moral and Epistemic Virtues, edited by Michael Brady and Duncan Pritchard (2003)

The Range of Pragmatism and the Limits of Philosophy, edited by Richard Shusterman (2004)

The Philosophical Challenge of September 11, edited by Tom Rockmore, Joseph Margolis, and Armen T. Marsoobian (2005)

Global Institutions and Responsibilities: Achieving Global Justice, edited by Christian Barry and Thomas W. Pogge (2005)

Genocide's Aftermath: Responsibility and Repair, edited by Claudia Card and Armen T. Marsoobian (2007)

Stem Cell Research: The Ethical Issues, edited by Lori Gruen, Laura Gravel, and Peter Singer (2007)

Cognitive Disability and Its Challenge to Moral Philosophy, edited by Eva Feder Kittay and Licia Carlson (2010)

Virtue and Vice, Moral and Epistemic, edited by Heather Battaly (2010)

Global Democracy and Exclusion, edited by Ronald Tinnevelt and Helder De Schutter (2010)

Putting Information First: Luciano Floridi and the Philosophy of Information, edited by Patrick Allo (2011)

The Pursuit of Philosophy: Some Cambridge Perspectives, edited by Alexis Papazoglou (2012)

Philosophical Engineering: Toward a Philosophy of the Web, edited by Harry Halpin and Alexandre Monnin (2014)

Philosophical Engineering

Toward a Philosophy of the Web

Edited by

Harry Halpin and Alexandre Monnin

WILEY Blackwell

First published as *Metaphilosophy* volume 43, no. 4 (July 2012), except for "Toward a Philosophy of the Web: Foundations and Open Problems," "Philosophy of the Web: Representation, Enaction, Collective Intelligence," "The Web as Ontology: Web Architecture Between Rest, Resources, and Rules," "Interview with Tim Berners-Lee," and "Afterword: Web Philosophy."

Registered Office
John Wiley & Sons Ltd, The Atrium, Southern Gate, Chichester, West Sussex, PO19 8SQ, UK

Editorial Offices
350 Main Street, Malden, MA 02148-5020, USA
9600 Garsington Road, Oxford, OX4 2DQ, UK
The Atrium, Southern Gate, Chichester, West Sussex, PO19 8SQ, UK

For details of our global editorial offices, for customer services, and for information about how to apply for permission to reuse the copyright material in this book please see our website at www.wiley.com/wiley-blackwell.

The rights of Harry Halpin and Alexandre Monnin to be identified as the authors of the editorial material in this work have been asserted in accordance with the UK Copyright, Designs and Patents Act 1988.

Library of Congress Cataloging-in-Publication Data applied for

Paperback ISBN: 978-1-118-70018-1

A catalogue record for this book is available from the British Library.

Cover image: The first page of Tim Berners-Lee's original proposal for the World Wide Web, March 1989. © 1989 CERN, reproduced with permission.
Cover design by Design Deluxe

Set in 10 on 11 pt Times by Toppan Best-set Premedia Limited
Printed in Malaysia by Ho Printing (M) Sdn Bhd

1   2014

CONTENTS

NOTES ON CONTRIBUTORS

Tim Berners-Lee invented the World Wide Web while at CERN in 1989. He is director of the World Wide Web Consortium, a Web standards organization he founded in 1994, which develops interoperable technologies to lead the Web to its full potential. He is a professor of engineering at the Laboratory for Computer Science and Artificial Intelligence at the Massachusetts Institute of Technology, and also a professor in the Electronics and Computer Science Department at the University of Southampton, UK.

Selmer Bringsjord is professor of logic and philosophy, computer science, cognitive science, and management and technology at Rensselaer Polytechnic Institute, the oldest technological university in the English-speaking world. He specializes in artificial intelligence and computational cognitive science, including the formal and philosophical foundations of both fields.

Andy Clark is professor of philosophy at the University of Edinburgh, and chair in logic and metaphysics. He previously taught at the Universities of Glasgow, Sussex, Washington (St. Louis), where he was director of the Philosophy/Neuroscience/Psychology Program, and Indiana. His extensive publications on embodied cognition, neural networking, and cognitive science made him one of the key figures in the field. These include the keystone book *Being There* as well as *Supersizing the Mind*.

Naveen Sundar Govindarajulu is a postdoctoral research associate at the Rensselaer AI & Reasoning Lab of Rensselaer Polytechnic Institute. His research continues work in building cognitively robust synthetic characters through formal modeling of self-consciousness in synthetic characters and building characters able to pass the mirror test for self-consciousness. His Ph.D. research overlaps with a Templeton Foundation-funded project, "Toward a Markedly Better Geography of Minds, Machines, and Math," to study and advance the mathematical frontiers of artificial intelligence research.

Harry Halpin is a visiting researcher at Institut de Recherche et d'Innovation du Centre Pompidou (PHILOWEB EU Marie Curie Fellowship). He is also a postdoctoral associate at Massachusetts Institute of Technology with the World Wide Web Consortium. He received his Ph.D. in Informatics from the University of Edinburgh under the supervision of Andy Clark and Henry Thompson. He published his thesis under the title *Social Semantics*. He is president of LEAP (LEAP Encryption Access Project), which works on providing open-source secure tools for activists.

Yuk Hui is a postdoctoral researcher at the Centre for Digital Cultures at Leuphana University, Lüneburg. He holds a Ph.D. in philosophy from Goldsmiths, University of London, and a B.Eng. in computer engineering from the University of Hong Kong. His Ph.D. thesis, "On the Existence of Digital Objects," proposes a new philosophical understanding of data and experience by bridging the works of Martin Heidegger, Edmund Husserl, and Gilbert Simondon.

Pierre Livet was formerly professor at the University of Franche-Comté and the University of Provence, and is now professor emeritus at the University of Aix-Marseille. His areas of research are the epistemology and ontology of social sciences (sociology and economics), the theory of action, the theory of emotions, and the epistemology of cognitive sciences. He is a member of CEPERC and IMÉRA at the University of Aix-Marseille.

Alexandre Monnin holds a Ph.D. in philosophy from Paris 1 Panthéon-Sorbonne, on the philosophy of the Web. He is associate researcher at Inria Sophia-Antipolis, where he co-initiated the francophone DBpedia project and the SemanticPedia platform, and co-chair of the W3C "Philosophy of the Web" Community Group. He was named one of the twenty-five experts of Etalab, the French government's open-data agency under the responsibility of the prime minister. He was previously visiting fellow at Internationales Kolleg für Kulturtechnikforschung undMedienphilosophie (Bahaus University, Weimar) and head of Web Research at Institut de Recherche et d'Innovation, Centre Georges Pompidou for three years.

Thomas W. Simpson is University Lecturer in Philosophy and Public Policy at the Blavatnik School of Government, University of Oxford. After completing his doctoral thesis entitled "Trust on the Internet" in the Faculty of Philosophy at the University of Cambridge, he took up a research fellowship at Sidney Sussex College, Cambridge. His research interests are in testimony and social epistemology, trust, applied ethics in the philosophy of technology, and the ethics of war.

Paul R. Smart is a senior research fellow in the School of Electronics and Computer Science at the University of Southampton. His main area of research is the Semantic Web, but he is also involved in research activities that sit at the interface of the network, Web, and cognitive sciences. He is currently involved in research that seeks to further our understanding of the effect that network-enabled technologies have on cognitive processes.

Johnny Hartz Søraker is assistant professor in the Department of Philosophy at the University of Twente, where he received his Ph.D. cum laude, on virtual worlds and their impact on quality of life. He has published extensively on topics within computer ethics and the philosophy of technology, and is currently working on a project that aims to synthesize the ethics of technology and positive psychology.

Petros Stefaneas is a lecturer in the Department of Mathematics at the National Technical University of Athens. He has studied at the University of Athens, the University of Oxford (in Programming Research Group), and the National Technical University of Athens. His research interests include logic and computation, formal methods, the Web and philosophy, the information society, and new media.

Bernard Stiegler, philosopher, is president of the Association Ars Industrialis, director of the Institut de Recherche et d'Innovation du Centre Pompidou, and professor at the University of London (Goldsmiths College). He is associated with the Technical University of Compiègne (UTC) and is a visiting professor at the University of Cambridge as well as ETH Zurich. Currently a member of the French national advisory board on digital matters (Conseil National du Numérique); he is the author of twenty-five books.

Michalis Vafopoulos is an adjunct professor at the National Technical University of Athens and previously taught in the Department of Cultural Informatics at the University of the Aegean. In 2009 he served as local chair of the First Web Science Conference and cofounded the master in Web science program at Aristotle University, Thessaloniki. In 2010 he introduced the official Web Science Subject Categorization. His research interests include Web economics, business, economic networks, open data, Web philosophy, and didactics.

Ioannis M. Vandoulakis is a lecturer in the History and Philosophy of Science at the Hellenic Open University. He holds a Ph.D. in the history of mathematics from Lomonosov Moscow State University and was research fellow at the Russian Academy of Science (Moscow) and CNRS (Paris). He has published on the history and philosophy of Greek mathematics, the history of mathematical logic, and the foundations of mathematics, including "Plato's Third Man Paradox: Its Logic and History."

Michael Wheeler is professor of philosophy at the University of Stirling. His primary research interests are in philosophy of science (especially cognitive science, psychology, biology, artificial intelligence, and artificial life) and philosophy of mind. He also works on Heidegger, and is particularly interested in developing philosophical ideas at the interface between the analytic and the continental traditions. His book *Reconstructing the Cognitive World: The Next Step* was published by MIT Press in 2005.

There are excellent philosophers of physics, philosophers of biology, philosophers of mathematics, and even of social science. I have never even heard anybody in the field described as a philosopher on engineering—as if there couldn't possibly be enough conceptual material of interest in engineering for a philosopher to specialize in. But this is changing, as more and more philosophers come to recognize that engineering harbors some of the deepest, most beautiful, most important thinking ever done.

—Daniel C. Dennett, *Darwin's Dangerous Idea: Evolution and the Meanings of Life* (1995), p. 120

We are not analyzing a world, we are building it. We are not experimental philosophers, we are philosophical engineers.

—Tim Berners-Lee, Message to W3C Technical Architecture Group mailing list (2003)

Computer scientists have ended up having to face all sorts of unabashedly metaphysical questions. . . . More recently, they have been taken up anew by network designers wrestling with the relations among identifiers, names, references, locations, handles, etc., on the World Wide Web.

—Brian Cantwell Smith, *On the Origin of Objects* (1995), pp. 44–45

CHAPTER 1

TOWARD A PHILOSOPHY OF THE WEB: FOUNDATIONS AND OPEN PROBLEMS

ALEXANDRE MONNIN AND HARRY HALPIN

Introduction

What is the philosophical foundation of the World Wide Web? Is it an open and distributed hypermedia system? Universal information space? How does the Web differ from the Internet? While the larger ecology of the Web has known many a revolution, its underlying architecture in contrast remains fairly stable. URIs (Uniform Resource Identifiers), protocols like HTTP (HyperText Transfer Protocol), and languages such as HTML (HyperText Markup Language) have constituted the carefully evolved building blocks of the Web for more than two decades. As the particular kind of computing embodied by the Web has displaced traditional proprietary client-side applications, the foundations of Web architecture and its relationship to wider computing needs to be clarified in order to determine the Web's roots and boundaries, as well as the historical reasons for its success and future developments. Crafting a philosophy of the Web is especially urgent, as debate is now opening over the relationship of the Web to platform computing on mobile devices and cloud computing.

The scope of the questions that the philosophy of the Web provokes is quite wide-ranging. These questions begin with the larger metaphilosophical issue of whether or not there are unifying principles underlying the architecture of the Web that justify the existence of a philosophy of the Web. Tim Berners-Lee, widely acclaimed as the inventor of the Web, has developed in his design notes various informal reflections over the central role of URIs (Uniform Resource *Identifiers*, previously *Locators*) as a universal naming system, a central topic in philosophy since at least the pioneering works of Barcan Marcus. URIs such as http://www.example.org/ identify anything on the Web, so the Web itself can be considered the space of all URIs. Thus, in brief, we would say that there is indeed at least one unifying principle to the architecture of the Web, that of URIs. The

Philosophical Engineering: Toward a Philosophy of the Web, First Edition. Edited by Harry Halpin and Alexandre Monnin. Chapters © 2014 The Authors except for Chapters 1, 2, 3, 12, and 13 (all © 2014 John Wiley & Sons, Ltd.). Book compilation © 2014 Blackwell Publishing Ltd and Metaphilosophy LLC. Published 2014 by Blackwell Publishing Ltd.

various architects of the Web, including Berners-Lee, made a number of critical design choices, such as creating a protocol-independent universal naming scheme in the form of URIs as well as other less well-known decisions, such as allowing links to URIs to not resolve (leading to the infamous "404 Not Found Message," a feature not allowed in previous hypertext systems) that—little to the knowledge of everyday users of the Web—do form a coherent system, albeit one that has not yet been explicated through a distinctively philosophical lens.

A critic could easily respond that there is no a priori reason any particular technology deserves its own philosophy. After all, there is no philosophy of automobiles or thermostats. Why would one privilege a philosophy of the Web over a philosophy of the Internet? These questions can be answered by looking at the nature of the design choices made in the formation of the Web: namely, in so far as the Web is based on URIs, the architecture of the Web exists on the level of naming and meaning, both of which are central to semantics and so are traditionally within the purview of philosophy. What the Web adds to the traditional philosophical study of natural language is both the *technically* engineered feat of a universalizing naming scheme in the form of URIs and the fact that such names can be accessed to return concrete bits and bytes, a distinctive feature of naming on the Web. However, the Web itself is agnostic over how the concrete low-level bits that compose something like a web page are transmitted across the network in response to an access request to a URI, as this is determined by protocols such as the Internet's TCP/IP (Transmission Control Protocol/Internet Protocol). Thus, the Web can be considered an abstract information space of names above the networking protocol layer, up to the point that it could have been (or could still be) built on top of another networking protocol layer (such as OSI [Open Systems Interconnection] or the "Future Internet"). Likewise, the Internet can also host applications other than the Web that do not use URIs, such as peer-to-peer file sharing or the Web's early rivals (the Gopher system, for instance). So in response to our critic, the Web does have its own architecture, and—unlike the case with automobiles and even the Internet—this architecture uncontroversially deals with philosophical concepts of naming and meaning, and this justifies the existence of a philosophy of the Web, at least insofar as names and meaning on the Web differ from natural language (or the philosophical way to conceptualize it!), a topic worthy of further exploration (Monnin 2012a).

The Web is not all protocols and naming schemes; it is also a wide-ranging transformation of our relationship to the wider world "out there," to the ontology of the world itself. It is precisely this engineering aspect that makes the philosophy of the Web differ qualitatively from traditional philosophy of language, where it has been assumed that natural language is (at least for philosophical purposes) stable and hence "natural." In contrast, the nature of the growth of both the Web and digital technologies undoubtedly calls into question the contemporary transformation of our entire form of life. Bringing scrutiny to bear on Wittgenstein's naturalistic concept of the "form of life," American sociologist Scott Lash takes into account the anthropological upheaval caused by the evolution of various mediums of thought on our technological forms of life (Lash 2002), a subject that has been abundantly discussed in the context of the Web (Halpin, Clark, and Wheeler 2014). Our main focus here, however, is less the future of humanity than that of philosophical research and philosophy itself. The architecture of the Web reveals a process of continuation and regrasping (which precisely needs to be properly assessed) of the most central of philosophical concepts: object, proper name, and ontology. On the Web, each concept of philosophy in its own way then gains a new existence as a technical artifact: objects turn into resources, proper names into URIs, ontology into Semantic Web ontologies.

Such a transition from philosophical concepts to technical objects isn't a one-way process and cannot remain without consequences for the original concepts that have been uprooted from their normal context, and accordingly this transition warrants careful examination. Do we philosophize today as we did in the past? With the same subject matter? Or in the same manner? Does it still make sense to locate oneself within established traditions, such as phenomenology and analytic philosophy, when their very own concepts freely cross these boundaries, and the real conversation is taken up elsewhere, using a language that only superficially seems identical to the one that preceded it? These kinds of questions have always been central to metaphilosophy, yet the advent of the Web—and so the philosophy of the Web—brings to these questions both a certain renewed importance and impetus. In the essays collected here, we bring together a number of authors who have offered some key contributions to this initial foray into the tentative realm of the philosophy of the Web. In order to guide philosophers through this nascent philosophical field, in the next section we delve deeper into the philosophical role of URIs and engineering as these two subjects serve as the twin foundations of the philosophy of the Web, and we then put each of the contributions in this collection within its philosophical context before reaching some tentative conclusions for next steps for the field.

1. URIs: "Artifactualization" of Proper Names

On the Web, the analogue of proper names is found in URIs, given by the standard IETF RFC 3986 to be "a simple and extensible means for identifying a resource," a definition in which resources are left crucially underdefined to be "whatever might be identified by a URI" (Berners-Lee,

Fielding, and Masinter 2005). URIs are everywhere: everything from mailto:harry@w3.org (for identifying an e-mail address of Harry Halpin) to http://whitehouse.gov (for identifying the page about the White House) qualifies as a URI. What quickly becomes apparent is that URIs are kinds of proper names for objects on the Web.

During the past fifteen years, philosophical discussions around the notion of a proper name have seamlessly followed in a business-as-usual manner, without any significant breakthrough. Yet during that same period, the architects of the Web have taken hold of the idea of proper names, and without purposefully altering its definition, have made naming the first supporting pillar of the Web, thus formulating an answer to the ages-old question of the relationship between words and things by combining in an original—and unintentional!—fashion the thoughts of Frege, Russell, Wittgenstein, and Kripke. For philosophy to take the URI, an engineered system of universal and accessible names, as a first-class philosophical citizen is then the first task of the philosophy of the Web.

While at first URIs may seem to be just a naming system for ordinary objects on the Web like e-mail addresses and web pages, the plan of Berners-Lee is to extend URIs as a naming scheme not just for the Web but for all reality—the Semantic Web will allow URIs to refer to literally anything, as "human beings, corporations, and bound books in a library can also be resources" (Berners-Lee, Fielding, and Masinter 2005). This totalizing vision of the Web is not without its own problems. In a striking debate between Berners-Lee and the well-known artificial intelligence researcher Patrick Hayes over URIs and their capacity to uniquely "identify" resources beyond web pages, Berners-Lee held that engineers decide how the protocol should work and that these decisions should determine the constraints of reference and identity, while Hayes replied that names have their possible referents determined only as traditionally understood by formal semantics, which he held engineers could not change but only had to obey (Halpin 2011). This duality can be interpreted as an opposition between a material and a formal a priori. Interestingly enough, recently, Hayes and other logicians such as Menzel have begun focusing on adopting principles from the Web into logical semantics itself, creating new kinds of logic for the Web (Menzel 2011). Unlike philosophical systems that reflect on the constraints of the world, the Web is a world-wide embodied technical artifact that therefore creates a whole new set of constraints. We suggest that they should be understood as a material a priori—in the Husserlian sense—grounded in history and technology.

Thus the Web, when it comes to its standards, breaks free from French philosopher Jules Vuillemin's definition of a philosophical system as built on the logical contradictions between major philosophical schools of thought (Vuillemin 2009). Yet the Web doesn't lead either to the collapse of the transcendental and the ontological into the empirical, a new kind of "technological monism" as suggested by Lash (2002). *Logical contradiction* is overcome not by *factual opposition* (two words that Vuillemin highlighted) but through an *artifactual composition*, associating through the mediation of the artifact the virtues of competing philosophical positions. As the functions of concepts become functionalities, it is becoming increasingly easier to make them coexist for the sake of a *tertium datur*, without having to give up on consistency (Sloterdijk 2001).

The material a priori of technical systems such as the Web is brought about by what we call "artifactualization" (Monnin 2009), a process where concepts become "embodied" in materiality—with lasting consequences, as the result trumps every expectation, being more than a mere projection of preexisting concepts (which would simply negate the minute details of the object considered). While such a process clearly predates the Web, we can from our present moment see within a single human lifetime the increasing speed at which it is taking place, and through which technical categories (often rooted in philosophical ones) are becoming increasingly dominant over their previously unquestioned "natural" and "logical" counterparts. At the same time, the process of having philosophical ideas take a concrete form via technology lends to them often radically new characteristics, transforming these very concepts in the process. Heidegger posited a filiation between technology and metaphysics, with technology realizing the Western metaphysical project by virtue of technology inscribing its categories directly into concrete matter. Yet if technology is grounded in metaphysics, it is not the result of a metaphysical movement or "destiny" (*Schicksals*), but a more mundane contingent historical process, full of surprises and novelties. For all these reasons, it must be acknowledged that the genealogy of the Web, as a digital information system, differs from traditional computation with regard both to the concepts at stake and to our relation to them. The scientific *ethos* is indeed being replaced by an engineering one, something Berners-Lee dubbed "philosophical engineering" (Halpin 2008)—and this difference even holds true with regard to the (mainly logical, thanks to the Curry-Howard correspondence) *ethos* of computer science itself.

As already mentioned, URIs form the principal pillar of Web architecture, so it shouldn't be surprising that they also constitute our gateway into the aforementioned problematic between engineering and philosophy. From its inception, the Web was conceived as a space of names, or "namespace," even if the historical journey to URIs led through a veritable waltz of hesitations as the engineers who built the Web tried to pin down standardized definitions to various naming schemes. The numerous Web and Internet standards around various kinds of names bear witness to that ambivalence: URL (Uniform Resource Locator), URN (Uniform Resource Name), and even URC (Uniform Resource Characteristic or Citation). Each of these acronyms matches a different conception of the

Web and modifies the way it constitutes a system. Eventually, the acronyms have slowly evolved over time to return to Berners-Lee's original vision of a URI: a "Universal" Resource Identifier for everything, from which follows naturally the ability of links to allow everything to be interconnected on the Web.

The notion of proper name as it prevails today is directly inherited from analytic philosophy, and more precisely, from the work of Saul Kripke; although other definitions may exist in philosophy, Ruth Barcan Marcus (Humphreys and Fetzer 1998) is clearly the one who launched this Kripkean tradition, and this strand of work eventually meant that "proper name" would become the key operating term for questions on reference, identity, and modality. It holds such a weight that it explains how fields as diverse as epistemology and ontology can be considered part of a larger story, that of a *science of reference*. This space of convergence was historically opened by the different theories of intentionality and objects from Brentano to Twardowski and Meinong, but it was to split post-Frege philosophy into two rival traditions, the analytical and the phenomenological—the latter sometimes considered "continental" from the analytical perspective (Benoist 2001). Ruth Barcan Marcus and Saul Kripke's works on proper names provide the apex for the analytic tradition, but what we see now on the Web is the URI as a proper name and technical object that reopens a space for reunification of these two divergent philosophical traditions within the philosophy of the Web as the problems around reference and naming migrate from philosophical systems (Vuillemin 2009)—in particular, the philosophy of language— toward *technical* and *artifactual* systems, asking for a complete shift of analysis.

A clear example of how URIs are transforming the analytic tradition's bedrock of logic has recently been pioneered by Patrick Hayes, known for his original quest to formalize common-sense knowledge in terms of first-order logic of artificial intelligence but also more recently deeply involved for several years in the development of the Semantic Web, the extension of the Web beyond documents into a generic knowledge representation language (Hayes 1979). As a logical foundation for the Semantic Web, Hayes has suggested the creation of "Blogic" (a contraction of "WeB logic," inspired by a similar contraction of "Web logs" to "blogs"), in which logical proper names, which possess no signification of their own outside their formally defined role in logic, would be replaced by dereferenceable URIs, which could in turn dereference logical sentences or even new interpretation functions not present in their original context.[1] Blogic would leverage the ability to use a name—in this case, a URI—to retrieve a "document," functionality that has played a critical part in the Web's

---

[1] See the talk "Blogic or Now What's in a Link?" by Patrick Hayes, online at http://videolectures.net/iswc09_hayes_blogic/.

success to this day, but outside hypertext and in the realm of semantics. With access mechanisms then possibly defining the semantics of proper names, the notion of reference on the Web cannot clearly choose between Wittgenstein (meaning determined by use), Russell (definite descriptions), and Kripke (rigid designators) for a theory of meaning. As a framework, the architecture of the Web composes with these conceptual positions: a user is free to give any kind of meaning to a URI, someone publishing a new URI may refer to it rigidly or with the help of a description for the Semantic Web, and what we access via that URI can also play a role in defining its meaning.

This mixture of the technical and the philosophical is found not only in the semantics of URIs but also in their governance (the latter having an impact on the former). URIs are not just free-floating names but assigned virtual territory controlled by bodies such as the Internet Assigned Numbers Authority (IANA) via domain registrars. While being proper names, URIs also have a particular legal and commercial status that does not clearly compare to proper names in philosophy, with perhaps only a vague analogue to the ability of organizations to copyright names. For websites, controlling a name, and thus which web pages can be accessed from it, is a source of immense power. According to Tim Berners-Lee, the ability to mint new URIs and link them with any other URI constitutes not only an essential linguistic function but also a fundamental freedom. Yet as URIs leave the field of semiotics, they undergo a change in nature as regards both the possibilities offered by a technology of naming and the limitations imposed by the legislation that governs bodies such as IANA. Once again, objects such as URIs or disciplines such as philosophy that seemed purely formal are gaining a newly found materiality, full of historical and even political contingencies around extremely concrete juridical and economic issues.

2. Denaturalizing Ontology: Philosophical Activity Redux

Another field in which the Web is rapidly causing massive conceptual tremors is the once forgotten philosophical realm of ontology. Given the long-lasting gap between a name and its object, the study of URIs on the Web naturally causes an intrepid philosopher of the Web to lean on work on names in philosophy of language, while with the study of ontology on the Web we return to the preponderance of the object. Generally considered a branch of metaphysics, ontology traditionally has generally been the study of the (often possible) existence of objects and their fundamental categorization and distinctions. Interestingly enough, engineering-inclined artificial intelligence researchers (the late John McCarthy being first among them) have also seized upon the word "ontology" over the past fifty years, making "ontology" their own term for purposes of creating knowledge representation languages, as exemplified by Gruber's famous engineering definition that "an *ontology* is a specification of a conceptualization" (Gruber 1993). While this definition may at first glance seem so vague to be totally useless, one should remember that Tim Berners-Lee also had notorious trouble defining precisely what a URI is, and this did not seem to prevent URIs from becoming central to the entire edifice of the Web. In fact, one would almost suspect that the utility of a term may somehow be related to the fact that it is underdefined—or perhaps more precisely, defined just enough to allow concrete engineering to reveal the inherent productivity of the term's concept. While the use of ontology by knowledge representation has become sidelined in philosophical circles by more clearly philosophical debates in artificial intelligence around embodiment, the move of Berners-Lee to create a Semantic Web that transforms the Web from a space of URIs for hypertext documents to a giant global knowledge representation language built on URIs has led to a renewal of interest in the engineering of ontologies as well. We suspect this computational turn in ontologies on the Web will in turn lead to a revival of the philosophical field of ontology.

As the shift from philosophical ontology to ontological engineering progresses, philosophers are gradually losing control over their own tools, even if they are not necessarily aware of it. What ensues is a real "proletarianization," as Bernard Stiegler (1998) puts it, and this process is smooth and passive, since philosophical activity goes on uninterrupted, as if nothing were amiss. Nonetheless, there are a number of unmistakable signs. Following the example of Barry Smith, some philosophers have already made their move explicit by rebranding themselves ontologists, as they are now working exclusively in the field of knowledge engineering. Might the conundrum of this technological life form be all about employing the concepts of philosophy in a new light while at the same time making the previous blissfully technologically unaware philosophical discipline obsolete? Although there are possibly some examples to illustrate such a strong point, a more reasonable response would be to answer this question with a little more subtlety by taking into account the precise nature of how the Web transforms ontology before tackling the wider question of how the Web transforms philosophy.

The ontological implications of the Web are deeply related to the concept of a resource harbored at the heart of Web architecture; for the philosophy of the Web this particular concept constitutes an opportunity to renew the question of ontology itself. As designated by a URI, a resource can be "anything at all," exactly as was the case for the hoary philosophical concept of the "object," which was the actual focus of the ontological tradition (as long as you trace the word back to its origins in the seventeenth century, more than twenty centuries after Aristotle's definition of the science of being). Consequently, it is not the sole business of philosophers, hidden far from the world at the back of some unidentified abode where they can hone their weapons alone to issue judgments and decide what "anything at all" really means. Everywhere, the gap is narrowing and the old privileges are in crisis, as Scott Lash (2002) has discussed. The new indexing tools and the contributive nature of the Web make it possible for anyone to tackle this issue—not only philosophers and engineers. Thus, the way in which the question will be asked relies on the Web itself. Nevertheless, in opposition to Lash's thesis, ontology is not suffering from being conditioned by the patterns of our technological way of living. Technology is the condition of the liberation of ontology, the "ontogonic" dimension of technology discussed by Bruno Bachimont (2010) and more recently with regard to the notion of philosophical engineering in Monnin (2012b) that draws on Pierre Livet's work presented in this collection (Livet 2014).

According to Livet's recent work on the ontology of the Web, we can chart the operations that allowed the emergence of the objects that in turn proved essential to conceptualize the Web's architecture and, subsequently, to clear the ontological horizon (Livet 2014). Far from consisting merely of epistemic processes, this work opens the door for an ontology of operations constitutive of an ontology of entities, which bit by bit refines itself as time goes by. The possibility to move back and forth, as the whole process unfurls, is not to be excluded, leading to new beginnings and thus leaving entirely open the question of the nature of the ultimate constituents of our technological cosmos.

Given the conceptual purity required by formal ontology, built as it is upon relationships of dependency and the application of mereology, and the materiality of devices as the place of a new technicized a priori, the time has come for a re-evaluation of the very notions of form and matter, through the filter of digital technologies in general and the Web in particular. In this regard, the research initiated several years ago on Ontology Design Patterns (Gangemi and Presutti 2009), which may appear to be limited only to the field of knowledge engineering, in fact has implications far beyond the boundaries of its original field. Incidentally, nothing prevents philosophers from trying to conceive their practice in a more collaborative fashion using a similar pattern analysis of their own activity, in order to gain a better view on the collective fine-grained ontological invariants that groups of philosophers share beyond the explicit debates through which philosophers normally distinguish themselves. Beyond this, the background against which these ontological patterns appear is formed by practices that, though they may produce certain apparently transhistorical regularities, are rooted in an historical context and therefore should not be "naturalized" prematurely.

In order to identify and qualify these invariants by taking into account that which supports and maintains them, one has to, so to speak, "denaturalize ontology"—and this slogan could serve as a synthesis of the entire philosophical research program we are suggesting here. To pretend, as is

often the case in analytic philosophy, that certain ontological constructions are simply pregiven would be a serious error, for everything has a
cost—one need only consider the works of Bruno Latour and Pierre Livet
to realize that (Latour 2001; Livet 2014). The key is determining how
technology opens an avenue into the historicization of ontology.

There are clear predecessors in either explicitly or implicitly building
ontologies into technology as well as having technology influence our
everyday ontology. In the fields of cognitive science and artificial intelligence, the question of the representation, formalization, and computation
of knowledge—as well as the more philosophically neglected approaches
centered on collective intelligence and human computation that partially
go beyond traditional philosophy of the mind due to their focus on human
intelligence's complementarity with the machine—have already produced
interesting leads in the wake of the work of Andy Clark and David
Chalmers on what they call "the Extended Mind Hypothesis" (Clark and
Chalmers 1998). We would like to extend the extended mind by renewing
metaphysics through a focus on the positive aspect of French linguist and
semiotician François Rastier's critique of cognitive science, which he
accused of "naturalizing metaphysics" (Rastier 2001).[2] Given the lessons
learned from the implicit metaphysics of cognitive science and artificial
intelligence, we cannot simply criticize or reject the ongoing exploitation
of the Web on an unprecedented scale that harnesses vast resources of
centuries of philosophical debates on language and knowledge. In order
to describe the paramount importance of the technical production of
media ranging from television to computers in the twentieth century,
Bernard Stiegler (1998) coined the expression "the machinic turning-point
of sensibility." With a slight shift of focus, we may talk of a *machinic* (or
perhaps better, *artifactual*) *turning-point of metaphysics* itself, an ongoing
deep modification of the meaning of metaphysics in philosophy.

At the present moment, the perspective given by the original architecture of the Web needs to be broadened as two problematics currently
intersect: (a) the *artificialization* of a growing number of particular domains
("natural" but also "formal" ones,[3] each of these two notions being traditionally contrasted with technics but now becoming technical) and (b)
the *artifactualization* of philosophical concepts in general as they are
imported into the realm of the digital—in particular under the guise of

---

[2] Of course, one of the necessary conditions to be spared the risk of unduly naturalizing
metaphysics is to make sure that no kind of prophetic and biologically inclined conception
of technology is allowed to thrive in parallel.

[3] The origin of this paradox could be found in Husserl's work, in which formal and material a prioris and ontologies are articulated. As always, this is only a starting point, and this
distinction is likely to be taken up and revised by other currents of thought. We think about
material cultures or authors whose works, although they differ in many ways, all put the
stress on the materiality of the mediums and operations of knowledge.

the dominant sociotechnical form of the twenty-first century, which is none other than the Web. It may even be argued that semiotic objects, such as philosophical concepts, already have features similar to those of technical objects (Halpin 2008),[4] in which case this latest round of digitization on the Web may rather appear tantamount to a *re-artifactualization*, provided that we do not ignore the (often overlooked) original ties of philosophy to technology.

Behind the distinction between the two aforementioned problematics stands an important issue: if these two dimensions are not clearly acknowledged, there is a risk that we will "naturalize" (a) without any real philosophical scrutiny the re-introduction (b) of some body of philosophical knowledge (or some unconscious philosophical legacy) while designing technical systems.[5] For all these reasons, the very practice of philosophy is transformed by having to take the material a priori and its technical categories as seriously as "natural" (synthetic) or "analytic" categories from biology or natural language. Philosophers then have to deal with engineered categories that may have a lasting effect in domains like the Web, not just as variants of categories that can be analytically understood but rather as concrete artifacts that can even transform analytic categories previously taken for granted. Ironically, the main challenge to analytic judgment is no longer what Quine called naturalization, but rather the ongoing artifactualization of which the Web is the historical exemplar par excellence (Livet 2014; Monnin 2013).

3. Open Problems of the Philosophy of the Web

Now that we have surveyed some of the core foundations of the nascent philosophy of the Web, we should be clear that we have only started to embark on this particular route, and the road ahead lies littered with open philosophical problems worthy of a tassel of theses. The Web was the brainchild not of Tim Berners-Lee as a lone individual but of a large and heterogeneous group of Web architects, ranging from Berners-Lee's compatriots involved in standards, such as Larry Masinter and Roy Fielding (Berners-Lee, Fielding, and Masinter 2005), to users who contribute

---

[4] On the articulation of semiotic and technical aspects, see also the work of Bruno Bachimont (2010).

[5] In this view, the warning is also useful against all ways of thinking that proceed step by step (two steps, to be precise), "computationalizing" or "informationalizing" the world first, in order to be subsequently entirely free to naturalize ontology, which is afterward considered a natural science of the universe (a tendency that historically dates back to the Neoplatonic philosophers). Opposing this way of thinking, Jean-Gabriel Ganascia advocates a computerized epistemology that is directly related to the cultural sciences, an epistemology that is well aware of the status of computers in the production of contemporary knowledge. In this respect, our point of view is quite close to his (Ganascia 2008).

content daily to the Web; the philosophy of the Web must likewise be a collective affair. We will provide brief summaries of some of the most pressing questions that face the philosophy of the Web, along with an overview of those contributors in this collection who have addressed them.

3.1. What Is the Relationship of the Philosophy of the Web to a More General Philosophy?

It goes almost without saying that every new discipline must build its foundations on earlier philosophical studies and situate itself consciously within a wider historical context, yet to do this correctly is one of the most difficult tasks for a philosopher. Case in point: one of the factors distinguishing the Web from earlier visions of an interconnected global information space, such as Vannevar Bush's Memex (Bush 1945), is that the Web is implemented on digital computers. This simple decision to stick to a digital medium had a deep impact on engineering matters, allowing as it does functionality from the copying and caching of pages used by search engines to the transformation of music and video by streaming media. Yet this almost obvious engineering decision in turn leads to a decidedly deep impact on metaphysical notions such as that of a resource (Berners-Lee, Fielding, and Masinter 2005). Yet in expanding the Web into the Semantic Web, where seemingly non-digital things are construed as resources, there still seems something self-evidently different between web pages and things themselves, as one can easily copy pages about the Eiffel Tower, due to their being digital, but not the Eiffel Tower itself. However, precisely explicating the philosophical difference between a page about the Eiffel Tower and the Eiffel Tower itself seems to always involve punting the question to an under-theorized notion of digitality. While there has been considerable philosophical debate over the nature of logic and computation, there has been little work on the wider notion of digitality. One of the prime tasks of the philosophy of the Web is to determine how the Web is engineered on top of robustly digital objects. This is not to say that the Web must remain digital forever—as the Web transforms into an Internet of Things and increasingly interacts with the analogue world, understanding digitality becomes more—not less!—important, as does situating this digital turn within wider currents in the analytic and phenomenological traditions. Yuk Hui's contribution "What Is a Digital Object?" provides a synopsis of his much larger foray into this field, and artfully combines Husserl's phenomenology with Simondon's understanding of technics to provide just such a theoretical foundation for the philosophy of the Web (Hui 2014).

3.2. Does the Web Radically Impact Metaphysics, Ontology, and Epistemology?

As we indicated earlier, it appears that the Web is having perhaps its most crucial role in philosophical realms that seem at first glance rather distant from high-speed technology. Taking on board the efforts of the Semantic Web to rework the logical foundations of ontology, Monnin (2014) argues that the main innovation of the architecture of the Web is definitely more ontological than technical. Raising the question "What do Web identifiers refer to and how?" he examines the answer provided by Web architects themselves. The investigation, in a way, is reminiscent of Quine's landmark paper "On What There Is," once it has been brought to the Web. Monnin's conclusion is that the Web articulates a deeply subtle view of objects on a global scale, theoretically as well as technically. In other words: the Web is an operative "ontology," as suggested in the title of his piece. Pierre Livet demonstrates in his contribution "Web Ontologies as Renewal of Classical Philosophical Ontology" how far the Web has taken us from classical ontological questions based on "natural" kinds to new dynamic and open-ended ontologies (Livet 2014). On a similar note, without a doubt the Web seems to be impacting the phenomenology of such fundamental metaphysical categories as space and time for ordinary users. Michalis Vafopoulos's contribution "Being, Space, and Time on the Web" is precisely such a retheorization of metaphysics on the Web, taking such fundamental aspects of the Web as the number of links in a page and reconceiving of this as a concept of space, with the time spent by users visiting a given resource as a concept of time, and then drawing a number of social and economic conclusions (Vafopoulos 2014). Moving from the world itself to our knowledge of the world, we find that one of most interesting phenomena brought about by the Web is the tendency of people to increasingly rely on search engines to answer their everyday questions. Precisely how ubiquitously search engine usage impacts classical conceptions of epistemological questions of knowledge and belief are tackled in the piece by Thomas Simpson, "Evaluating Google as an Epistemic Tool" (Simpson 2014). These epistemological questions are also not purely theoretical, as obviously the use of search engines in everyday environments, political debate in wikis and (micro)blogs, and the possible transformation of the university system itself by massive Web-mediated online courses all merit serious attention, and how we understand these issues is directly influenced by our position on whether or not access to knowledge on the Web counts as belief or even knowledge.

3.3. Can Human Cognition and Intelligence Genuinely Be Extended by the Web?

Questions about epistemology naturally lead to questions about whether or not the Web is changing our conception of humanity, and to questions about how the philosophy of the Web interacts with other empirically informed philosophical questions around neuroscience and cognitive science. In particular, the question of whether or not human cognition is genuinely extended by the Web appears rather naturally: In a not-so-distant common scenario when humans are wearing Google-enabled goggles that allow them to almost instantly and seamlessly access the Web, would one give the Web some of the cognitive credit for problem solving? This very example, in terms of a special pair of glasses rotating blocks in the classic game of Tetris, is brought up as one of the motivating examples of the classic Extended Mind Hypothesis of Chalmers and Clark (Clark and Chalmers 1998), and now this example is coming very close to reality on the Web. However, the precise conditions of what constitutes the mark of the cognitive, and under what conditions the Web actually counts as part of an extended mind, are still very much a topic of debate. The first essay in this collection, "Philosophy of the Web: Representation, Enaction, Collective Intelligence," outlines connections between the philosophy of the Web and what has been termed "'4E' (embodied, embedded, enactive, extended) cognition (Halpin, Clark, and Wheeler 2014). In this overview, Halpin, Clark, and Wheeler note how concepts from cognitive science such as representation and enaction must be re-thought through in the light of the Web's status as a readily accessible externalized public cognitive resource. They end with broaching the topic of how the Web brings forth the possibility of not just an extended mind (Clark and Chalmers 1998) but also a new type of massively distributed collective intelligence that has yet to be properly studied via the lens of philosophy. In his contribution, "The Web-Extended Mind," Paul Smart mounts an argument for why the Web should be considered part of the extended mind, and then proceeds to show how engineering developments such as Berners-Lee's Semantic Web may have certain design characteristics that could more tightly bind the mind to the Web in the future (Smart 2014). This naturally implies revisiting the question of intelligence. Interestingly enough, in the early days of the Internet, the motivating vision was one of collective intelligence, in which machines augmented rather than replaced human intelligence. This vision was a direct rival of artificial intelligence, which hoped to implement human-level intelligence in machines. Surprisingly, a thorough investigation of the philosophical assumptions and differences of both collective and artificial intelligence has yet to be written, an investigation that has become increasingly urgent, as the Semantic Web itself is often criticized as a mere repeat of classical artificial intelligence. A first important step has been taken by Selmer Bringsjord and Naveen Sundar Govindarajulu in their contribution "Given the Web, What Is Intelligence, Really?" in which they convincingly argue that even if the Semantic Web did become a reality, it would lack the reasoning capacity of humans that intelligence requires (Bringsjord and Govindarajulu 2014). How this particular debate over the transformation of intelligence on the Web plays out in the future will no doubt have yet unforeseen ramifications, just as the original quest for artificial intelligence radically revised the traditional pre-computational philosophy of the mind.

3.4. Does the Web Alter Our Domain-Specific Practices in a Manner That Demands a New Qualitative Analysis?

Given our order of presentation, one of the central questions of the philosophy of the Web would seem to be how the new engineering-inspired revolution in the rarefied air of philosophy will have its tremors felt in various specialized domains. Quite the reverse seems true: the impact of the Web has most powerfully been noticed in its empirical effects on almost impossibly heterogeneous domains, ranging from online recommendation systems in e-commerce to the near instantaneous spread of news globally via microblogging. Indeed, the primary difficulty of the philosophy of the Web lies precisely in the difficulty inherent in tracing how such a diverse range of complex induced effects could form a coherent philosophical system, one that may have explanatory and even predictive power. Thus, detailed domain-specific studies of how the Web impacts particular domains of practice are critically part of the philosophy of the Web. Few areas can be considered seemingly more remote from mundane engineering considerations than proof-proving in mathematics, yet in their remarkable contribution, "The Web as a Tool for Proving," Petros Stefaneas and Ioannis Vandoulakis demonstrate the nature of the radical impact of the Web on this most formal and theoretical of domains (Stefaneas and Vandoulakis 2014). To move in the reverse direction, the communication and ubiquitous accessibility of the Web may alter our notion of embodiment. Nowhere is this more powerfully demonstrated than in the multiplayer Web-mediated virtual worlds, whose numbers seem to be growing everyday. While currently only a small part of the Web, it is very possible that such deeply immersive and even "three-dimensional" environments may come of age soon and become an important part of the future of the Web. Johnny Hartz Søraker engages with these environments in "Virtual Worlds and Their Challenge to Philosophy: Understanding the 'Intravirtual' and the 'Extravirtual,'" where he carefully compares and contrasts the kinds of actions possible in these worlds (Søraker 2014). This leads one to think that fundamentally the Cartesian distinction between the "real world" and the "virtual world" may indeed be far more complex than initially conceived, positing a problematic that may end up being just as important for the philosophy of the Web as the mind–body problem is for the philosophy of mind. Far from being purely academic, these debates over how the Web interacts with our daily life are already stirring upheaval in how we understand our own notion of privacy and identity, and will soon perhaps even take political center stage.

3.5. The Future of the Philosophy of the Web

In this collection, we have endeavoured to take into full account both the engineering aspect and the wider philosophical ramifications of the Web.

So we are pleased to feature new pieces from both Tim Berners-Lee and Bernard Stiegler on philosophical engineering and the philosophy of the Web. We are privileged to include with his permission the text of an interview we did with Berners-Lee, widely acclaimed as the inventor of the Web (Halpin and Monnin 2014). In this wide-ranging interview, Berners-Lee reveals why he coined the term "philosophical engineering" and offers his thoughts on the future of the philosophy of the Web. We end the collection with an afterword entitled "Web Philosophy" by Bernard Stiegler, one of France's preeminent philosophers of technology, who uncovers the potential and the "shadows" at the heart of the digital enlightenment's new political philosophy (Stiegler 2014). Stiegler attempts to place "philosophical engineering" in proper context by uncovering its lineage from Archimedes to the Web by way of philosophers such as Plato, Husserl, and Derrida.

4. Conclusion

Ultimately, the philosophy of the Web has just begun, and its future is far from certain: the impact of the Web may ultimately be as transformative as that of natural language, or perhaps it will be superseded within a short time by some truly distinct technological development. Regardless, just as the study of artificial intelligence provoked genuine philosophical inquiry into the nature of mind and intelligence, the philosophy of the Web will at least—we hope!—provoke the taking of engineered artifacts such as the Web that impinge on areas traditionally the province of philosophy as first-class subjects of concerted inquiry by philosophers. The Web has obviously benefitted from previous encounters with philosophically informed engineers, although this point alone would deserve a more ample treatment (Shadbolt 2007).

All these engineering-related activities on the Web can easily be described, if one is willing to alter Clausewitz's adage, as "philosophy continued through other means." Most of these means, though, are far from being completely exogenic. It is only through a redoubling of awareness over these new technological mediations, studied and developed by "philosophical engineering," that philosophers may have the opportunity to extend their categories by submitting them to the evaluation of the non-human technical artifacts, now regular constituents of our world. The stake here is a change of course that has nothing to do with just switching one given a priori (or *epistēmē*) for another. Ian Hacking (2002) sharply underlined that such notions are far too massive. We believe that those interested in the philosophy of the Web are not simply wearing a new pair of conceptual glasses; the world itself has changed, for it is composed no longer of canonical entities but of utterly new ones that differ fundamentally from their predecessors. Enriched with the new details bequeathed to it by technology, the world is asking for a redefinition of its ontological cartography (perhaps an "ontography"), even if it implies that we should broaden our philosophical focus to encompass all the agents responsible for the shift to the Web (engineers, languages, standardization committees, documents, search engines, policymakers, and the like). No empirical metaphysics is entitled to define the "nature," patterns, or limits of the agents that can impact the philosophy of the Web without argumentation and clarity. What is at stake, the determination of the "collective," the cosmos we live in, requires that we stay on the verge of philosophy itself, on the very technical spot where new objects[6] are spreading, and once these objects are brought back to their philosophical womb, they will certainly foster mutation within philosophy. We consider the Web to be endowed with this capacity more than any other technical apparatus, and that the duty of the philosophy of the Web is to set ourselves upon the task.

As we have seen, this task leads to a number of open questions, and we hope that the contributions to this collection have put forward some of these questions as clearly as possible, in order to highlight theoretical as well as technological issues—and even social and political matters— that will set the philosophy of the Web in motion toward the interdisciplinary point of view necessary to adequately address these problems. In the context of the Web more than ever, neither philosophy nor engineering can escape its practical consequences by dodging issues that are relevant, and even vital, to how the architects of the Web have in the past engineered and will in the future engineer the Web itself, as it increasingly becomes the primary medium of knowledge and communication. Again, the traffic is not only one-way between philosophy and the Web, it is a dynamic feedback cycle: as the Web itself is mutating as a medium, we can consider humans and objects of knowledge to be condemned to mutate in turn. Historically, philosophy is a discipline descended from the alphabet and the book. Under the influence of what has been called by French historian of language Sylvain Auroux a "third revolution of grammatization" (Auroux 1994) catalyzed by digital technologies and the Web, what turn philosophy takes remains to be seen.

Richard Sennett's motto in a recent book appears to be "doing is thinking" (Sennett 2008); once concepts have been artifactualized (and, as a consequence, externalized), thinking is also doing, and so in the end, a matter of design. In this regard, we need to reject Marx's Eleventh Thesis on Feuerbach: in the era of the Web, interpreting the world is already changing it. This holds true especially when the art of interpretation and theorizing is serving the purpose of the creation of new forms of technology that harness the power of the Web via building on and creating

---

[6] Several of these new objects have been minutely examined by other disciplines, and we may want to borrow from them. One meaningful example is the analysis of standards as an essential tool to understand the Web's architecture.

Web standards. These innovations target a reality yet to come, but a reality that we can already conceive of, as opposed to a pre-existing reality. The matter of engineering the Web is rapidly transforming into the matter of engineering the world. To paraphrase Saussure, the philosopher and the engineer are both challenged to acquire a clear view of what it is they are doing. For the philosophy of the Web and philosophy itself, these are the stakes.

Afterword

A few words are necessary in order to provide some context for the recent development of the philosophy of the Web that led to this collection, and to provide a call for those interested to join in shaping the future of the philosophy of the Web. The term "philosophy of the Web" was first coined by Halpin (2008), while the central importance of artifactualization in the philosophy of the Web was first explicated by Monnin (2009). Monnin organized the first Web and Philosophy symposium in 2010 at La Sorbonne. Subsequent editions of this symposium were organized jointly by Halpin and Monnin in 2011 at Thessaloniki to coincide with the conference on the Theory and Philosophy of Artificial Intelligence and at Lyon in 2012 as a workshop at the International World Wide Web Conference. Several of the contributions to this collection are extended versions of presentations given at these events, and we would like to thank all of those who have participated in the discussion so far. Again, the philosophy of the Web is not the static product of a single individual (or even two!), but a collective endeavour like the Web itself, whose scope and power widens the more that people are involved. To engage in future events and public debate, join the W3C Philosophy of the Web Community Group: http://www.w3.org/community/philoweb/.

References

Auroux, Sylvain. 1994. *La révolution technologique de la grammatisation.* Liège: Mardaga.
Bachimont, Bruno. 2010. *Le sens de la technique: Le numerique et le calcul.* Paris: Encre Marine.
Benoist, Jocelyn. 2001. *Représentations sans objet: Aux origines de la phénoménologie et de la philosophie analytique.* Paris: P.U.F.
Berners-Lee, Tim, Roy Fielding, and Larry Masinter. 2005. "IETF RFC 3986—Uniform Resource Identifier (URI): Generic Syntax" Available at https://tools.ietf.org/html/rfc3986.
Bringsjord, Selmer, and Naveen Sundar Govindarajulu. 2014. "Given the Web, What Is Intelligence, Really?" Included in this collection.

Bush, Vannevar. 1945. "As We May Think." *Atlantic Magazine*, vol. 7. Available at http://www.theatlantic.com/magazine/toc/1945/07.

Clark, Andy, and David Chalmers. 1998. "The Extended Mind." *Analysis* 58:10–23.

Ganascia, Jean. 2008. "'In Silico' Experiments: Towards a Computerized Epistemology." *APA Newsletter on Computers and Philosophy* 7, no. 2:11–15.

Gangemi, Aldo, and Valentina Presutti. 2009. "Ontology Design Patterns." In *Handbook on Ontologies*, second edition, edited by S. Staab and R. Studer, 221–43. Berlin: Springer.

Gruber, Thomas. 1993. "A Translation Approach to Portable Ontology Specifications." Available at http://tomgruber.org/writing/ontolingua-kaj-1993.htm.

Hacking, Ian. 2002. *Historical Ontology*. Cambridge, Mass.: Harvard University Press.

Halpin, Harry. 2008. "Philosophical Engineering: Towards a Philosophy of the Web." *APA Newsletter on Philosophy and Computers* 7, no. 2:5–11.

———. 2011. "Sense and Reference on the Web." *Minds and Machines* 21, no. 2:153–78.

Halpin, Harry, Andy Clark, and Michael Wheeler. 2014. "Philosophy of the Web: Representation, Enaction, Collective Intelligence." Included in this collection.

Halpin, Harry, and Alexandre Monnin. 2014. "Interview with Tim Berners-Lee." Included in this collection.

Hayes, Patrick. 1979. "The Naive Physics Manifesto." In *Expert Systems in the Micro-Electronic Age*, edited by D. Michie, 242–70. Edinburgh: Edinburgh University Press.

Hui, Yuk. 2014. "What Is a Digital Object?" Included in this collection.

Humphreys, Paul, and James Fetzer. 1998. *The New Theory of Reference: Kripke, Marcus, and Its Origins*. Berlin: Springer.

Lash, Scott. 2002. *Critique of Information*. London: Sage.

Latour, Bruno. 2001. *Pasteur: Guerre et paix des microbes, suivi de Irré-ductions*. Paris: La Découverte.

Livet, Pierre. 2014. "Web Ontologies as Renewal of Classical Ontology." Included in this collection.

Menzel, Christopher. 2011. "Knowledge Representation, the World Wide Web, and the Evolution of Logic." *Synthese* 182, no. 2:269–95.

Monnin, Alexandre. 2009. "Artifactualization: Introducing a New Concept." In *Proceedings of Interface09, 1st National Symposium for Humanities and Technology*. Southampton: University of Southampton.

———. 2012a. "L'ingénierie philosophique comme design ontologique: Retour sur l'émergence de la 'ressource.'" *Réel-Virtuel* 3, Archéologie des nouvelles technologies. Available at http://reelvirtuel.univ-paris1.fr/index.php?/revue-en-ligne/3-monnin/2/.

————. 2012b. "The Artifactualization of Reference and 'Substances' on the Web: Why (HTTP) URIs Do Not (Always) Refer nor Resources Hold by Themselves." *American Philosophical Association Newsletter on Philosophy and Computers* 11, no. 2:11–9.

————. 2014. "The Web as Ontology." Included in this collection.

Rastier, François. 2001. *Sémantique et recherches cognitives.* Paris: P.U.F.

Sennett, Richard. 2008. *The Craftsman.* New Haven: Yale University Press.

Shadbolt, Nigel. 2007. "Philosophical Engineering." *Words and Intelligence II: Essays in Honor of Yorik Wilks,* edited by Khurshid Ahmad, Christopher Brewster, and Mark Stevenson, 195–207, Text, Speech, and Language Technology series 36. Dordrecht: Springer.

Simpson, Thomas W. 2014. "Evaluating Google as an Epistemic Tool." Included in this collection.

Sloterdijk, Peter. 2001. "Die Domestikation des Seins: Für eine Verdeutlichung der Lichtung." In *Nicht gerettet: Versuche nach Heidegger,* edited by P. Sloterdijk, 142–234. Berlin: Suhrkamp.

Smart, Paul R. 2014. "The Web-Extended Mind." Included in this collection.

Søraker, Johnny Hartz. 2014. "Virtual Worlds and Their Challenge to Philosophy: Understanding the 'Intravirtual' and the 'Extravirtual.'" Included in this collection.

Stefaneas, Petros, and Ioannis M. Vandoulakis. 2014. "The Web as a Tool for Proving." Included in this collection.

Stiegler, Bernard. 1998. *Time and Technics: The Fault of Epimetheus,* Volume 1. Palo Alto, Calif.: Stanford University Press.

————. 2014. "Afterword: Web Philosophy." Included in this collection.

Vafopoulos, Michalis. 2014. "Being, Space, and Time on the Web." Included in this collection.

Vuillemin, Jules. 2009. *What Are Philosophical Systems?* Cambridge: Cambridge University Press.

CHAPTER 2

PHILOSOPHY OF THE WEB: REPRESENTATION, ENACTION, COLLECTIVE INTELLIGENCE

HARRY HALPIN, ANDY CLARK, AND MICHAEL WHEELER

Introduction

There is an emerging vision of the human mind as essentially a social organ apt to make extensive and transformative use of whatever forms of local and global scaffolding other agents and technologies provide. In an increasingly wired and networked world, our very nature as cognitive beings is gradually changing.

The World Wide Web is a remarkable triumph of incremental computational engineering. In the wake of its technological success and the structural change it has effected on human social organization, Web designers and researchers are being forced to confront a range of foundational issues with clear philosophical dimensions. These include old philosophical issues in modern guises—issues concerning knowledge, identity, and trust, as well as new questions raised by the increasingly complex ways in which the Web is embedded in the larger world. Such new questions concern, for example, the character and status of Web objects, such as websites and mash-ups, the understanding of authorship within new collaborative and collective creative ensembles, and the relation between, on the one hand, the structure and functioning of the Web and, on the other, the strengths and weaknesses of basic biological cognition.

1. Is Philosophy Part of Web Science?

Sir Tim Berners-Lee, the inventor of the Web, has challenged philosophy to contribute to the future construction of the Web, calling the architects behind the Web "philosophical engineers" (Berners-Lee 2003). Our hypothesis is that profitable philosophical engagement with the Web will be achieved through the lens of contemporary debates in philosophy of mind and cognitive science over what is sometimes called *4E (embodied,*

Philosophical Engineering: Toward a Philosophy of the Web, First Edition. Edited by Harry Halpin and Alexandre Monnin. Chapters © 2014 The Authors except for Chapters 1, 2, 3, 12, and 13 (all © 2014 John Wiley & Sons, Ltd.). Book compilation © 2014 Blackwell Publishing Ltd and Metaphilosophy LLC. Published 2014 by Blackwell Publishing Ltd.

embedded, enactive, extended) cognition. Indeed, it is this area of philosophy that has attracted the most interest from leading thinkers about the Web, although to date there are very few examples of philosophers of 4E cognition engaging directly with the Web. As we explain below, a number of critical and fundamental questions that Web researchers are beginning to confront are intertwined with issues at the forefront of recent work on 4E cognition. Importantly, however, the intellectual traffic here is not one-way. A carefully crafted philosophy of the Web will not merely draw on the aforementioned philosophical debates, it will make important contributions to them. The study of human–Web couplings provides a powerful way to pursue several unresolved and controversial contemporary philosophical issues. In this way, practical interests in the design and use of Web-based technologies dovetail with foundational questions concerning the nature of minds and persons.

This new philosophy of the Web should go beyond much "new media studies" work on the Web by engaging directly with certain pressing scientific and engineering concerns faced by Web architects. What was missing from this earlier work was the productive engagement with the relevant engineering community that the present project will foster. Much of the fault of this lies on the shoulders of philosophy, however (Dreyfus 2001). The debate over relevance sensitivity to be found in the 4E philosophical literature has seldom been connected to the performance of search engines or the Web more generally (Wheeler 2005). Similar comments might be made about enactivist thinking. By contrast, the extended mind hypothesis is often accepted as almost intuitively obvious in Web circles, even though in philosophical circles it is the subject of ongoing critical debate (Adams and Aizawa 2001). The established interest in the extended mind by the Web community is indicated by, for example, work of computer scientists and psychologists on applying the extended mind hypothesis to the Web (Smart et al. 2010). Moreover, the extended mind hypothesis was advanced as a possible foundation for a philosophy of the Web in a 2008 exchange between Halpin and Wheeler in the *American Philosophical Association Newsletter on Philosophy and Computers* (Halpin 2008; Wheeler 2008). Building upon this previous work, we can outline three linked themes that a philosophy of the Web based on empirical work and cognitive science should address: the problem of representations on the Web, enactive search, and collective intelligence.

2. Representations and the Web

Cognitive science and artificial intelligence (AI) have traditionally appealed to internal and neutrally realised representations to explain intelligence. But exactly how the notion of representation should be understood, and the extent to which neurally located representations are necessary for intelligence, have long been vexed issues, especially in the philosophical literature. The Web provides a new impetus for investigation into the notion of representation, because the advent of the Web has seen an explosion of novel external representations (e.g., hypertext web pages) that, through complex, iterated interactions with human users, enable intelligent information retrieval, complex commercial activity, and social communication and coordination. We must explore the relationship between these two seemingly different representational contributions to intelligent activity, asking whether there is a single account of representation that applies to both of them.

These are theoretical questions with practical implications. The Semantic Web (also known as Linked Data) is a project launched by Berners-Lee and deployed by august bodies, including the U.K. government (e.g., in the recent initiative to release government data) and the BBC (Berners-Lee, Hendler, and Lassila 2001). Its goal is to build large-scale structured knowledge representation systems using the Web. Accordingly, it has been identified by some computer scientists and philosophers as an attempt to restart the project of classical AI (Floridi 2009). If this is right, the Semantic Web will ultimately face the problems that, according to some critiques, plagued its intellectual predecessor (Clark 1997). One such problem is the Frame Problem, the recalcitrant difficulty of determining, in a wholly mechanistic manner, which items of information from a huge memory store are relevant (and which are irrelevant), and how retrieved items of information should be updated, within and across changing contexts of activity (McCarthy and Hayes 1969). Influenced by phenomenological philosophers such as Heidegger (who stressed embeddedness) and Merleau-Ponty (who in addition stressed embodiment), thinkers such as Dreyfus have argued that the root cause of such problems is the assumption that the mechanistic processes at work are representation guided (Dreyfus 1979). Perhaps anti-representationalist-embodied cognition has itself been taken too far? Alternative remedies depend on a reconceived, Web-friendly notion of representation, and identify the implications for how the Web may meet the challenge of relevance sensitivity.

3. Enactive Search

Currently, access to representations on the Web is mediated through search engines such as Google. One of the keys to the practical success of such search engines is that they use massive amounts of statistics gathered from user actions and user choices on the Web to dynamically adapt their algorithms to find appropriate content, and thus to grapple with the issues of relevance and context highlighted by the frame problem. The more information a search algorithm has about the conduct and interests of users, the better its adaptation, a trend accelerated by "Web 2.0" technologies, such as social networking and collaborative tagging.

We hypothesize that the complex adaptive dynamics of such statistics-driven user-action-based searches may be illuminated using the philosophical concept of enaction (roughly, the idea of "laying down a path in walking," that is, of actions that change the world in ways that feed and structure those very actions, either now or at some future time [Varela, Thompson, and Rosch 1993]). Although much of the existing work on enaction has focused on the biological individual, an enactive paradigm can also be applied to the collective effects of our use and navigation of the Web. From this theoretical standpoint, we must investigate how search engines highlight a number of key issues in the way we think about minds, persons, and collective endeavours. These range from a kind of "quantification" of Wittgenstein's maxim "meaning is use" as a form of statistical language processing, an intellectual legacy dating back to the early search engines produced in Cambridge in the 1960s, to the empowering and disturbing (in about equal measure) vision of the not-too-distant future described by Google's CEO Eric Schmidt, in which Google is connected "straight to your brain" (Wilks 2005).[1]

4. Cognitive Extension and Collective Intelligence

According to the extended mind hypothesis as presented by Clark and Chalmers (1998), cognitive processes are not always confined within the boundaries of skin and skull. Under certain conditions in which biological brains and bio-external scaffoldings work together as integrated processing ensembles, cognition may extend into the world. The nature of the Web opens up this controversial philosophical issue in a distinctive and distinctively problematic way. Perhaps external representations on the Web, when integrated appropriately into the processes that govern an agent's behaviour, may count as parts of that agent's cognitive architecture. But now assume that multiple individuals are able to access the same external representation, such as a Google map, and that they can update it in near-real time. Here, it seems, more than one person may deserve cognitive credit for, and have cognitive ownership of, a representation that augments their own individual intelligence (Halpin 2008). What is still up for grabs is exactly which biotechnological couplings yield genuine extensions of cognition, rather than merely novel supporting environmental structures in which internally constituted cognition may function.

Finally, an analysis of the social dimension of Web-enabled distributed intelligence will help us to understand our increasing dependence on the

[1] See http://techcrunch.com/2009/09/03/google-ceo-eric-schmidt-on-the-future-of-search-connect-it-straight-to-your-brain/.

Web, not only for information gathering but also for socially coordinating action (via tools like social networking and microblogging). An adequate account of such collective intelligence (or the intelligence of collectives) on the Web will revisit existing theories of distributed cognition in the light of the case for the Web-extended mind (Hutchins 1995). A central challenge will be to analyze the conditions under which users trust, by responding unreflectively and uncritically to, the collectively maintained information retrieved from the Web. Plausibly (think of information retrieved from internal memory), such trust is a necessary condition for the external representations involved to qualify as cognitive extensions. From Wikipedia to Google, this ability to trust information on the Web is one of the most pressing problems facing the Web today (O'Hara and Hall 2008), so a detailed philosophical analysis of this topic promises to have a significant impact on the practice of Web engineers designing these systems.

5. From the Extended Mind to the Web

In order to hone some of these points, we put forward a speculative extension to the extended mind hypothesis that shows how the Web may affect the argument. As put forward by Wheeler, "online intelligence is generated through complex causal interaction in an extended brain-body-environment system" (Wheeler 2005, 12). We can press this point to make room for an active role of representations in general, and for the Web in particular, "an active externalism, based on the active role of the environment in driving cognitive processes" (Clark 1997, 6). For example, a representation can be stored in the memory "inside" the head of an agent in some neural state, but it can just as easily be stored outside in a map. The debate over the existence of internal representations is an empirical debate best left to neuroscientific work. What is less up for debate, however, seems to be that representations at least exist externally for particular agents. Finding those representational neural states is difficult, but let us not deny the existence of maps!

In the original argument for the extended mind hypothesis, Clark and Chalmers introduce us to Otto, a man with an impaired memory who navigates through his life via the use of maps and other such notes in his notebook (Clark and Chalmers 1998). Otto wants to navigate to the Metropolitan Museum of Modern Art in New York City from his house in Brooklyn, but to do so with his impaired memory he needs a map.[2] In order to arrive at the museum, Otto needs a map whose components are in some correspondence with the world he must navigate in order to get

---

[2] In fact, many of us would need a map even without an impaired memory, which points to how widespread this phenomenon is.

to the museum. Otto can find a map to the Museum of Modern Art that exists for the precise purpose of assisting individuals to navigate their way to the museum. It is hard to deny that a map is representational in the sense we have presented above, as it is a representation whose target is the various streets on the way to the museum. The map is just an external representation in the environment of Otto, and can drive the cognitive processes of Otto in a fashion similar to the way that classical AI assumed internal representations in Otto's head did. Clark and Chalmers point out that if external factors are driving the process, then they deserve some of the credit: "If, as we confront some task, a part of the world functions as a process which, were it done in the head, we would have no hesitation in recognizing as part of the cognitive process, then that part of the world is (so we claim) part of the cognitive process" (Clark and Chalmers 1998, 8).

Humans have difficulty maintaining any sort of coherent memory in the face of the vagaries of their dynamic environment. Technology allows this weakness to be turned into a strength, for the consequence is that humans who have a way to maintain an external memory capable of holding representations, even if such a memory is technically outside their biological skin, will likely have an evolutionary advantage. The entire progress of media forms, from speech to writing, from tablets to papyrus, from digital memory in personal computers to the Internet, can be considered the progress of this external memory, as detailed elsewhere (Logan 2000). Equally important is not just the ability to retrieve a representation from a perfect (or at least "more perfect") memory but that the representation can be accessed by more than one individual at once.

Imagine the world to be inhabited by multiple individuals that can access the same representation. In almost all the original examples that Clark and Chalmers (1998) use in the extended mind thesis, they deploy a single person sitting in front of a computer screen. A more intuitive example would be two people using the Internet to share a single representation. One could imagine Otto trying to find his way to the Museum of Modern Art, and instead of a notebook having a personal digital assistant with access to a map on the Web. Likewise, someone we might call Inga can have access to the exact same map via her personal digital assistant. Since both Otto and Inga are sharing the exact same representation and are both using it in the same manner, they can be said to be sharing at least some of the same cognitive state, due to the fact that their individual cognitive states are causally dependent on accessing the same representation. This representation is the "same" precisely because of the perfect, digital memory of the computer. Unlike the lone digital computer user of old, however, what the Web specializes in is allowing *everybody* to access the same set of representations.

The value of external representations comes with their accessibility, for an external representation that is not accessible when it is needed cannot be used to enable online intelligence. It is precisely in order to solve this problem that Tim Berners-Lee proposed a World Wide Web as a universal information space (Berners-Lee 1994). The primary advantage of the Web is that every representation has a unique name, a URI (such as http://www.example.org).[3] The Web allows each representation to be accessed when needed by using its unique name. This, combined with the fact that since the representations are digital and (at least can be) communicated in a lossless fashion, allows multiple simultaneous accessing of the exact same representation. Since the Web is a universal space of digital representations, two or more individuals can share the same representation simultaneously. Due to the extended mind hypothesis, two or more individuals can then, because of simultaneous access, share some of the same cognitive state.

6. The Web as Collective Intelligence

Much as computation has not remained static, neither has the Web. The Web, as originally conceived by its creators and early users, was just a collection of documents connected by hyperlinks, albeit in a universal information space. The documents were mostly static, being written and maintained by individuals. Although new pages and links could be added without resort to a centralized registry, the content of the Web was for the vast majority of users not content that they actually created and added to in any meaningful manner. Within the past few years, a combination of easy-to-use interfaces for creating content and a large number of websites that prioritize the social and collaborative creation of content by ordinary users have taken off, leading to the phenomenon known as Web 2.0, literally the next generation of the Web.[4] This transition from the Web of static hyperlinked web pages to a more interactive and collaborative medium is more accurately described as a transition from a "Web of Documents" to a "Social Web" (Hoshka 1998). Paradigmatic examples of easy-to-use interfaces would be Google Maps (or even Google Earth), while a paradigmatic example of socially generated content would be Wikipedia. Furthermore, these websites are now increasingly being woven into the fabric of the everyday life of more and more people. How many people feel that their intelligence is increased when they have immediate access to a search engine on the Web, a massive encyclopaedia available at a few seconds' notice?

---

[3] Originally, the "Universal Resource Identifier" (Berners-Lee 1994), now a "Uniform Resource Identifier," as given in an updated specification (Berners-Lee, Fielding, and Masinter 2005).

[4] A term originally coined by Tim O'Reilly for a conference to describe the next generation of the Web.

The Social Web, then, presents an interesting twist on the extended mind hypothesis extension that we presented earlier. Again, Otto is using a web page in his mobile phone to find his way to the Museum of Modern Art. While our previous example had Otto using the Web as ordinary Web users did years ago, simply downloading some directions and following them, we now add a twist. Imagine not only that Inga and Otto are using a map-producing website that allows users to add annotations and corrections, a sort of wiki of maps. Inga, noticing that the main entrance to the Museum of Modern Art is temporarily closed due to construction and so the entrance has been moved one block, adds this annotation to the map, correcting an error as to where the entrance of the museum should be. This correction is propagated at speeds very close to real time back to the central database behind the website. Otto is running a few minutes behind Inga, and thanks to her correction to the map being propagated to his map on his personal digital assistant, he can correctly navigate to the new entrance a block away. This (near) real-time updating of the representation is crucial for Otto's success. Given his memory issues, Otto would otherwise have walked right into the closed construction area around the old entrance to the museum and been rather confused. This active manipulation with updating of an external representation lets Inga and Otto possess some form of dynamically changing collective cognitive state. Furthermore, they can use their ability to update this shared external representation to influence each other for their greater collective success. In this manner, the external representation is clearly social, and the cognitive credit must be spread across not only multiple people but also the representation they use in common to successfully accomplish their behaviour. Clark and Chalmers's original paper states: "What about socially extended cognition? Could my mental states be partly constituted by the states of other thinkers? We see no reason why not, in principle" (Clark and Chalmers 1998, 17).

This leads us back full circle to the Web. For example, the collective editing of Wikipedia and its increasing use allow its representations to be increasingly part of the cognitive system of many people. As representations on the Social Web are updated by growing numbers of people, each representation is increasingly brought into tighter coupling with both its targets. As each representation involved in this process of use and updating is brought into an increasingly speedy cognitive updating with more and more individuals, the representations on the Web are brought into tighter and tighter coupling with what its users formerly considered their individual intelligence, so leading to the phenomenon widely known as collective intelligence. Indeed, there are now problems as simple as navigating down the street or organizing a social event that many today would find difficult without access to an interactive mapping Web service or a social networking website. As users contribute more and more content, the collective content of these web pages becomes increasingly difficult to track down to individuals. Some of these Web-based tools for collective intelligence have no way to track down the original individual author, while others like Wikipedia have sophisticated mechanisms in place to track individual contributions.

However, as long as the contribution that the collectively built web page makes is the sum of more than an individual effort, the credit must be given to the collective assemblage, not the individual author. From the standpoint of the user of the representation, the credit must be given not just to the creator of the content but also to the very technological infrastructure—ranging from the hardware of high-speed fibre optics and wireless routers to the software of protocol design and Web server code—that enables the content of the collectively created website to be delivered when it is needed. The credit for successfully creating and deploying the cognitive scaffolding is more collective than originally thought.

7. Conclusion

Having philosophers seriously move their research programmes to the nature of the Web will doubtless cause a paradigmatic shift in the debate over cognition and the extended mind, and thus more generally in the relationship between philosophy and the Web. A successful philosophy of the Web depends on taking an approach to the philosophical questions that remain grounded in the science and technology of the Web, including detailed rigorous inspection of empirical work we have not had the space to delve into here. It should be clear, however, that a careful analysis of a wide interdisciplinary literature is necessary, a literature that extends beyond the traditional grounds of cognitive science and into studies of online communities, human–computer interaction, information retrieval, hypertext, and the Semantic Web. Although we have not answered all the questions that a philosophy of the Web should answer in order to provide answers to outstanding questions from the philosophy of mind and language, we have at least made a map of the territory for future research. Philosophy may be part of Web science, after all.

References

Adams, Frederick, and Kenneth Aizawa. 2001. "The Bounds of Cognition." *Philosophical Psychology* 14, no. 1:43–64.

Berners-Lee, Tim. 1994. "IETF RFC 1630—Universal Resource Identifier (URI)." http://www.ietf.org/rfc/rfc1630.txt.

———. 2003. "Message on www-tag@w3.org List." http://lists.w3.org/Archives/Public/www-tag/2003Jul/0158.html.

Berners-Lee, Tim, Roy Fielding, and Larry Masinter. 2005. "IETF RFC 3986—Uniform Resource Identifier (URI): Generic Syntax." http://www.ietf.org/rfc/rfc3986.txt.

Berners-Lee, Tim, James Hendler, and Ora Lassila. 2001. "The Semantic Web." *Scientific American* 284, no. 5:35–43.
Clark, Andy. 1997. *Being There: Putting Brain, Body, and World Together Again*. Cambridge, Mass.: MIT Press.
Clark, Andy, and David Chalmers. 1998. "The Extended Mind." *Analysis* 58, no. 1:7–19.
Dreyfus, Hubert. 1979. *What Computers Still Can't Do: A Critique of Artificial Reason*. Cambridge, Mass.: MIT Press.
———. 2001. *On the Internet*. New York: Routledge.
Floridi, Luciano. 2009. "Web 2.0 vs. the Semantic Web: A Philosophical Assessment." *Episteme* 6, no. 1:25–37.
Halpin, Harry. 2008. "Philosophical Engineering: Towards a Philosophy of the Web." *APA Newsletter on Philosophy and Computers* 7, no. 2:5–11.
Hoshka, Philipp. 1998. "CSCW Research at GMD-FIT: From Basic Groupware to the Social Web." *ACM SIGGROUP Bulletin* 19, no. 2:5–9.
Hutchins, Edwin. 1995. *Cognition in the Wild*. Cambridge, Mass.: MIT Press.
Logan, Robert. 2000. *The Sixth Language: Learning a Living in the Internet Age*. Toronto: Stoddart.
McCarthy, John, and Patrick J. Hayes. 1969. "Some Philosophical Problems from the Standpoint of Artificial Intelligence." In *Machine Intelligence*, edited by Bernard Meltzer and Donald Michie, 4:463–502. Edinburgh: Edinburgh University Press.
O'Hara, Kieron, and Wendy Hall. 2008. "Trust on the Web: Some Web Science Research Challenges." *Universitat Oberta de Catalunya Papers: E-Journal on the Knowledge Society*, no. 7.
Smart, Paul R., Paula Engelbrecht, Dave Braines, Mike Strub, and Cheryl Giammanci. 2010. "The Network-Extended Mind." In *Network Science for Military Coalition Operations: Information Extraction and Interaction*, edited by Dinesh Verma, 191–236. Hershey, Penn.: IGI Global.
Varela, Francesco, Evan Thompson, and Eleanor Rosch. 1993. *The Embodied Mind: Cognitive Science and Human Experience*. Cambridge, Mass.: MIT Press.
Wheeler, Michael. 2005. *Reconstructing the Cognitive World: The Next Step*. Cambridge, Mass.: MIT Press.
———. 2008. "The Fourth Way: A Comment on Halpin's 'Philosophical Engineering.'" *APA Newsletter on Philosophy and Computers* 8, no. 1:9–12.
Wilks, Yorick. 2005. "A Personal Memoir: Margaret Masterman (1910–1986)." In *Language, Cohesion, and Form*, edited by Margaret Masterman, 1–18. Cambridge: Cambridge University Press.

CHAPTER 3

THE WEB AS ONTOLOGY: WEB ARCHITECTURE BETWEEN REST, RESOURCES, AND RULES

ALEXANDRE MONNIN

1. Introduction

Few researchers have dared to theorize the basics of the Web. Those who did characteristically borrowed concepts from their favorite philosophers or philosophical schools. Interestingly, one can illustrate this trend from both sides of the spectrum, from deconstructionism to analytic philosophy. According to one of the most well-known hypertext theorists, Georges Landow (see Delany and Landow 1991; Landow 1991, 1994, 1997, 2006), hypertexts—the Web being assimilated to one of them—enact what was first described by "deconstruction" and French theory from the 1960s (Barthes, Foucault, Derrida, Deleuze, and so on). As both these theories and hypertexts "converge," the technical artifact somehow becomes the striking *incarnation* of a set of preexisting concepts that deeply redefined the roles of the author and the reader, writing, textuality, and so forth.

In a distant philosophical universe, Nigel Shadbolt (2007), professor of artificial intelligence (AI) at the University of Southampton and an open data advocate, adopts a very similar stance while taking inspiration from a different—even antagonistic—tradition: namely, the analytic one. According to Shadbolt, the endeavor that motivated the works of philosophers such as Frege, Wittgenstein (the "early" one), Russell, and the Vienna Circle—undisputedly the fathers of the analytic tradition—was to shed light on the very notion of meaning. Their collective endeavor is broadly taken to have led to the definition of meaning used within the field of computing. In this narrative, the Semantic Web is the latest and most grandiose episode in a long tradition that goes back to Aristotle, matured in AI, and finds its apex with the Semantic Web. Rooted in a solid realistic view of objects, it purports to represent their intrinsic qualities in the vocabularies composing them and their properties and consisting of terms whose meanings correspond to objective divisions of reality.

Philosophical Engineering: Toward a Philosophy of the Web, First Edition. Edited by Harry Halpin and Alexandre Monnin. Chapters © 2014 The Authors except for Chapters 1, 2, 3, 12, and 13 (all © 2014 John Wiley & Sons, Ltd.). Book compilation © 2014 Blackwell Publishing Ltd and Metaphilosophy LLC. Published 2014 by Blackwell Publishing Ltd.

Finally, other authors have explicitly tried to analyze the new digital reality (so-called cyberspace) from an ontological perspective. David Koepsell in particular, opposing the extravagant claims regarding the fundamentally different metaphysical status of the digital realm, undertakes to reassess the mundane character of "retrieving" a Web page:

> Web pages are just another form of software. Again, they consist of data in the form of bits which reside on some storage medium. Just as with my word processor, my web page resides in a specific place and occupies a certain space on a hard drvie [sic] in Amherst, New York. When you "point" your browser to http://wings.buffalo.edu/~koepsell, you are sending a message across the Internet which instructs my web page's host computer (a Unix machine at the university of Buffalo) to send a copy of the contents of my personal directory, specifically, a HTML file called "index.html," to your computer. That file is copied into your computer's memory and "viewed" by your browser. The version you view disappears from your computer's memory when you no longer view it, or if cached, when your cache is cleaned. You may also choose to save my web page to your hard drive in which case you will have a copy of my index.html file. My index.html file remains, throughout the browsing and afterward, intact and fixed. (Koepsell 2003, 126–27)

All (or so it seems) the bits and pieces, the concrete inner workings that make up the plumbing behind such a simple action, are analyzed against a backdrop of available notions and thus made all the less mysterious.

Yet each of the three previous cases suffers from a common mistake. By taking for granted without any justification some concepts, they commit what Brian Cantwell Smith called an "inscription error," thus designating the ontological presuppositions superimposed on a given domain (i.e., the Web). Is the Web a hypertext? There are very good reasons to doubt that it is. Is it simply the continuation of the inquiry pursued by analytic philosophy and AI on the nature of meaning? Again, if there is an ontological issue to be raised, it might not be first and foremost with regard to computer ontologies. Finally, though Koepsell's scenario certainly rings a bell with any Web user, in fact, what it describes, what we might call the default view of the Web, is by and large *wrong*.

Furthermore, behind the aforementioned inscription error lies a tendency to forget the true character of technical systems and objects. While it is extremely tempting to think of the Web as the *realization* of preexisting concepts, and the new digital environment as one that gives new technical flesh to venerable ideas, this view entirely misses the point of what it is to be a technical artifact: more than a mere speculum in which philosophical ideas get reflected. To borrow an expression from Antoine Hennion and Bruno Latour (1993), such a view betrays the signature of "antifetishism," with the risk of simply losing sight of the objects.

In order to avoid both inscription errors and antifetishism, it was decided to stick to two principles. The first comes from Smith (1998), who calls it the "principle of irreduction"—with a clear nod to Bruno Latour. According to this principle, no concept or presupposition should a priori be given preeminence. Our second viaticum comes from none other than Latour himself. It consists in *following the actors themselves*, so as to be able to map out their "experimental metaphysics." Interestingly, very few researchers have deigned to take into account the work of dozens of Web architects in standardization bodies like the Internet Engineering Task Force (IETF) or the World Wide Web Consortium (W3C) before devising their own theories; even fewer have always avoided the danger of substituting preexisting concepts for the reality at hand in a careful analysis of the Web.

2. A Tale of Two Philosophies: URIs Between Proper Names and REST

To understand the evolution of the Web from a simple project to the global platform we now know, one has to pay attention to the evolution of one of its core elements, its naming system. Only then will the portrait of what these names identify slowly start to emerge. Yet this is a troubled story going through multiple stages, from the first papers published by Tim Berners-Lee around 1992 on UDIs (Uniform Document Identifiers [Berners-Lee, Groff, and Cailliau 1992]), the drafts of the first standards published at IETF on URIs (Uniform Resource Identifiers; RFC 1630 [Berners-Lee 1994]), the first real standards, published later the same year, after the creation of the W3C as part of the first wave of standardization of the Web, when URIs where sundered in URLs (conceived as addresses of documents; RFC 1738 [Berners-Lee, Masinter, and McCahill 1994]) and URNs (names of objects; RFC 1737 [Sollins and Masinter 1994]), to the modern understanding/implementation that revolves, once again, around URIs (RFC 2396 [Berners-Lee et al. 1998] and RFC 3986 [Berners-Lee, Fielding, and Masinter 2005]).

Each of these recommendations embodies and enacts a different understanding of the Web. The usual story weaves the tale of a Web of documents, the height of hypertext technologies, which would eventually become a Web of objects, the Semantic Web or Web of Data. This is what prompted the so-called "Web identity crisis" (see Clark 2002, 2003a, 2003b), a controversy that was unleashed at the turn of the century when people began to wonder whether it would be possible to use the Web to identify not only accessible "documents" but any kinds of "objects."

2.1. The Web Identity Crisis

The most visible symptom of the crisis is without a doubt the issue called HTTP-Range 14 (see Berners-Lee 2002; Fielding 2005). The HTTP-Range

TABLE 3.1. The HTTP-Range 14 (Lewis 2007)

Code	Result	Indication
200 ("OK")	HTTP representation	Information or non-information resource
303 ("See other")	URI	Any kind of resource
4XX/5XX	Error message	Impossible to infer anything about the nature of the resource

14 issue was an attempt to derive a technical criterion meant to distinguish between URIs that identify "information resources" (digital documents) from "non-information resources" (objects). The resolution of the issue was to use the code sent by a server to its client in HTTP headers to infer the nature of the resources identified by a URI (see table 3.1).

As is apparent from table 3.1, the technical result of the HTTP-Range 14 issue is that it is impossible to devise such a criterion: a URI that identifies Tim Berners-Lee could dereference content about him just as well as a URI that identifies a *page* about Tim Berners-Lee. Technically, there would be no difference. The same goes with redirection: it is not clear whether a URI that identifies "the Bible" and then redirects to one particular edition of the Book (such as the King James Version) would identify either a retrievable document or an abstract object in the first place. That is in fact why the decision that followed the HTTP-Range 14 issue proved to be a normative one, leading to the promotion of good practices, such as the 303 redirection, stating that for "inaccessible resources," one should redirect to accessible resources through a 303 redirection, using a second URI, served by a 200 HTTP code.

The conjunction of two elements explains this decision. First, the information/non-information distinction is rooted in the opposition, found in early proposals like UDIs, between objects and documents. It explained why URIs were sundered into URLs and URNs as soon as the standardization of Web identifiers was addressed by the W3C. The other factor had to deal with the importance inference engines acquired in the context of the Semantic Web, as those were not apt to make or ignore distinctions between a thing and its "representation," which are drawn fairly easily by human beings.

2.2. The Descriptivist Versus Rigidist Controversy

Beyond these local discussions, another reason is to be found in the inscription error made by the actors themselves in relation to the theories often associated with URIs, beyond the architecture of the Web. This second controversy is tightly related to the Web identity crisis. Holding URIs to be tantamount to philosophical proper names, it opposed an updated version of Bertrand Russell's descriptivist theory, championed by Patrick Hayes, and a position somehow akin to Saul Kripke's rigid designation. But each theory led to the same pitfall.[1] The extensionalist conception of objects drawing from model theory leaves beyond its scope the work of objectification that is necessary to deal with objects just as much as the rigidist position it opposes.

Indeed, in what seems to be the first discussion on proper names, dating back to 1962, Ruth Barcan Marcus's work in analytic philosophy on issues of identity and modality identified proper names with meaningless "tags" (Marcus 1995, 32–34). Marcus imagined that those tags would be used to pin down objects, as in a dictionary (she later admitted that this was a mistake and that what she had in mind was more encyclopedia-like, as Kripke himself noticed (Kripke 1980, 101). As if objects were "already there," just waiting for their Adam to be picked out and receive an individual tag, so that we could use this encyclopedia to determine for every object its name(s) in order to assess its (necessary) identity.

Marcus herself separated this problem from that of identifying the aforementioned objects, which is an epistemic one. In a sense, the Semantic Web, as conceived by Shadbolt, would appear to fulfill her project and truly "operationalize" philosophical ideas. Yet as evidenced by the lack of universal ontology (and without anyone taking this prospect seriously), the Semantic Web tends to experimentally demonstrate the contrary. Without this epistemic activity, there is no ontology to be found, as there is no means of determining what the ontological furniture of the world is. Only identifiers from this perspective, not objects, support relations of identity, after a world of "ready-made objects," to quote Hilary Putnam (2005), has first been hypothesized. That's the paradox of philosophical proper names, and taking them as the right tools to understand how URIs function although they were originally devised to answer questions of identity, logic, and language, whereas people on the Web are confronted by epistemic and ontological issues.

More generally, what is at stake here is the role of philosophy itself as a formal space where all the metaphysical distinctions would be already mapped, as in Jules Vuillemin's attempt (Vuillemin 2009) at formalizing philosophical systems (Vuillemin himself insisted on the importance of proper names for his perspective). Does philosophy purport to deliver concepts usable at any time and within any context because they exhaust every (logical? metaphysical?) possibility; or is it, as a discipline, capable of giving room to what Jacques Derrida (!) called "local thought events" (see Janicaud 2005, 124), thanks to which "actors themselves may locally change the metaphysics of the world," following Adrian Cussins' beautiful formula (Cussins 2001)?

---

[1] This is something I explore in Monnin 2013. For another pioneering account, see Halpin 2009.

After all, why would concepts from the philosophy of language constitute the best tool available to shed light on the problems that the architects of the Web are facing? Especially once these problems have been displaced from here to there? Not to mention the fact that there is barely any consensus on the very problems philosophers are supposed to solve, without even mentioning those of adjacent disciplines (such as linguistics, where philosophy's proper names are taken to be no more than mere artifacts that are nowhere to be found in ordinary language).

2.3. Back to REST

Contrary to what the focus on the Web identity crisis might indicate, the actual architectural style of the Web, known as REST (for *Re*presentational *S*tate *T*ransfer) comes up with a very different answer to the questions raised earlier. REST is the result of the work undertaken by Roy Fielding for his Ph.D. dissertation (Fielding 2000). Fielding had been tasked by Tim Berners-Lee to elaborate on the design philosophy behind HTTP and the Web. Fielding's mission, at the time, was to understand what made the Web so special, by going so far as to contrast the result of his investigation with the actual implementation of the Web. In other words, to be *faithful enough to the Web to go so far as to transcend its implementation.* That is why, in his answer to an e-mail questioning whether REST, which contains the principles behind the HTTP protocol, logically precedes it, Fielding points out that the answer to that question cannot rely on a merely "logical" chicken-and-egg distinction:

> Question: Logically, REST really had to predate HTTP 1.1 in order for HTTP 1.1 to be so RESTful.
> No?
> Answer (R.T. Fielding): No. That is more of a philosophical question than a logical one.
> HTTP/1.1 is a specific architecture that, to the extent I succeeded in applying REST-based design, allows people to deploy RESTful network-based applications in a mostly efficient way, within the constraints imposed by legacy implementations. The design principles certainly predated HTTP, most of them were already applied to the HTTP/1.0 family, and I chose which constraints to apply during the pre-proposal process of HTTP/1.1, yet HTTP/1.1 was finished long before I had the available time to write down the entire model in a form that other people could understand. All of my products are developed iteratively, so what you see as a chicken and egg problem is more like a dinosaur-to-chicken evolution than anything so cut and dried as the conceptual form pre-existing the form. HTTP as we know it today is just as dependent on the conceptual notion of REST as the definition of REST is dependent on what I wanted HTTP to be today. (Fielding 2006)

The W3C was created in 1994 in order to ensure the technical governance of the Web. Right after its creation, a first wave of standardization was launched. Among the first standards to be written were those related to Web identifiers, known first as UDIs and later URIs. URIs, as already mentioned, were then split into URLs and URNs, each with different kinds of referents, the former's being thought of as ever-changing accessible documents and the latter's as stable objects outside the Web. Yet this didn't seem to work out, because people wanted to access contents about any kind of "object" on the Web, using HTTP instead of ad hoc protocols designed and maintained by URN scheme owners, such as library organizations, for instance. For that reason, in 1997 and 1999, major revisions were brought to the standards, which included the fusion of URLs and URNs into URIs and a new version of the HTTP protocol. Fielding was among other things the main editor of the HTTP 1.1 protocol, including the current 1999 version (Fielding et al. 1997, 1999), and URIs (in 1998 and then again in 2005), and the cocreator of the Apache Foundation (which designed the dominant server software on the Web). With REST, one can see him participating in the establishment of a coherent Web, from servers, to the transfer protocol between them and their clients, to the underlying naming system that powers the entire construct.

The work on REST began around 1995, at a time when the difficulties surrounding URLs and URNs became more obvious by the day. Fielding made his doctoral dissertation widely available online in 2000. He also published an article written with his dissertation supervisor (Fielding and Taylor 2002). These constitute all the primary sources available on REST save for a few blog posts Fielding published years later (especially Fielding [2008]).

REST is often seen as a method for building Web services competing with SOAP (Simple Object Access Protocol), CORBA, Web Services, and other RPC protocols. Viewed thus, however, its real significance is completely lost. REST is more precisely described as what Latour calls a "re-representation" (Latour 2005, 566–67), a document whose purpose is to reinterpret a number of other technical documents (RFCs) and the artifact they describe (the HTTP protocol and URIs in particular). The work on REST had an immediate impact on the way standards were rewritten in 1997 and 1998 under Fielding's guidance. Despite not being a standard, it nevertheless acted as a "meta-investment of forms," to extend Laurent Thévenot's work on the investment of forms (Thévenot 1984) one step further, in order to make standards themselves more generic, stable, and interoperable (which is already the purpose of basic standardization). It is as much a reinterpretation of a technical reality that precedes it as a way to devise, discover, and/or ascribe (the frontiers between those terms are blurred) new distinctions that proved immensely useful (see Pierre Livet's contribution in this volume and Monnin 2013, parts 3 and 4, for a more thorough discussion on *distinguishing* as an ontogonic activity).

2.4. Resources as Shadows Symbolized Through Functions in REST

REST articulates a very original view of what's "on" the Web. According to Berners-Lee, URIs were not addresses, unlike UNIX paths. The difference between the two is mainly explained with regard to two different kinds of variations that I call "synchronic" and "diachronic."

Synchronic variations originate from functionalities, such as "content negotiation" (also known as "conneg"), a feature of the HTTP protocol that forbids taking one file stored on a server as what is being referred to by a URI, since the content served can vary according to the customization of the client's request, *making it impossible to functionally relate URI with one file on a server* (an assertion that holds from the inception of the Web, *pace* Koepsell). Furthermore, even "static" Web pages (which never really existed—a good example against this very idea would be the use of counters that changed their "pages" every time they were being accessed by a client) could be considered mash-ups, as they contained external URIs of images embedded in HTML elements, thereby distributing the sources from which such "pages" were generated. Such principles date back to 1991–1993, with System 33, a Xerox system demonstrated to Tim Berners-Lee by Larry Masinter at PARC in the early 1990s (Putz 1993; Putz, Weiser, and Demers 1993).

The other kind of variations, *diachronic variations*, are more widely acknowledged than the synchronic ones. It is well known that pages "evolve." Lacking any versioning system, the Web does not self-archive those modifications by adding a new identifier every time they occur. Yet "pages" somehow remain *the same*. Such paradoxical duality explains why URIs were at some point replaced by both URLs and URNs. At the time, it was thought that content varied too much on the Web to allow for stable reference, leaving this task to URNs, though URNs were devoid of any access function, at least through the HTTP protocol.[2] The dichotomy performed was a classic "on the Web"/"outside the Web" one, which was later abandoned thanks to REST—though this is something that still hasn't always been properly assessed. Another reason has to do with the fact that URLs are not addresses, or should not be treated as such, otherwise lots of problems arise because the *local* level of the database directories when it reflects the way URLs are written and constantly rewritten is no longer distinguished from the *global* level exposed to users, where stability is paramount; a necessity that is made clear through modules such as "URL rewriting" in Apache servers that help to manage both levels.

The solution advocated in REST (in a nutshell, without exhausting REST's significance for the Web) is a very elegant one. It states that instead of files, documents, objects, and so forth, what is being referred

---

[2] By default, everything on the Web will either change or disappear. Everything that has been published must be tended to. The idea that the so-called Web 1.0 was a Web of long-lasting, not-dynamic documents is by and large a phantasy.

to by a URI is a "resource." Here's the definition given in REST (by Fielding and Taylor): "A resource R is a temporally varying membership function $M_R(t)$, which for time t maps to a set of entities, or values, which are equivalent. The values in the set may be resource representations and/ or resource identifiers. A resource can map to the empty set, which allows references to be made to a concept before any realization of that concept exists—notion that was foreign to most hypertext systems prior to the Web" (Fielding and Taylor 2002, 125).

Resources are compared to concepts or even "shadows" insofar they are not material, unlike "representations," those encoded messages that are served when a URI is dereferenced. Whence the paradox, central to the Web and its architecture, described by Fielding and Taylor in a paragraph judiciously entitled "Manipulating Shadows":

> 7.1.2 Manipulating Shadows. Defining resource such that a URI identifies a concept rather than a document leaves us with another question: how does a user access, manipulate, or transfer a concept such that they can get something useful when a hypertext link is selected? REST answers that question by defining the things that are manipulated to be representations of the identified resource, rather than the resource itself. An origin server maintains a mapping from resource identifiers to the set of representations corresponding to each resource. A resource is therefore manipulated by transferring representations through the generic interface defined by the resource identifier. (Fielding and Taylor 2002, 135)

One interesting consequence of this paradox is that unlike other information systems, where pointers are for instance linked to memory addresses, resources are truly abstract. Hence, what is digital, the representation, is also material (as in "not virtual"). If there is anything virtual, in the philosophical sense of the word, as opposed to actual, as Deleuze reminded us, then it is *not* digital. In addition, we come to understand, thanks to REST, that the issue at stake is not so much "what there is *on* the Web" but rather "what there is," a question that is asked anew *thanks to the Web*.[3]

3. From References to Referentialization

3.1. Resources as Rules

In order to address these difficulties, one should decide against favoring philosophical concepts, such as proper names (understood as an *explanans*) to shed some light on those that were now stabilized in the

---

[3] Among other things, there are digital objects that have specific properties, being nonrivalrous goods, for instance (see Hui and Vafopoulos in this volume). Though ontologically challenging in their own way, they still differ from resources. A thorough analysis of the Web must account for both the latter and their representations.

current state of the architecture of the Web (resources, URIs, HTTP representations, etc.). We have already paid twice for the concepts we received from the actors we were studying: first by refusing to follow them when they appealed to the philosophy of language, then by following what they had "practically" done.

There is yet another distinction that calls for an explanation in REST: the idea that resources can be further articulated by *states* and *representations of these states* (whence the acronym REST itself: what is *Transferred* on the Web is a *REpresentational State* of a resource). We thus end up with three constitutive elements: the *resource* identified by the URI, the *state* entered at a given time, and the concrete, accessible *representation of that state*. I would like to argue that this threefold distinction leads to a comparison with Wittgenstein's concept of a rule, the resource being the equivalent of the rule; the state the result of rule following; and the representation a symbolic (or technical) representation of the latter.

The concept of a rule sheds light on a central difficulty for Web architecture. REST indeed specifies that a resource is always abstract, as opposed to its representations. Nevertheless, to argue that my resource is a chair requires associating the physical properties of a chair with the abstract properties that Web architecture attributes to the resource. As in Edward Zalta's (2003) take on fictional entities—whose relation to rules is more obvious than chairs!—where some properties are either *encoded* (being named Sherlock Holmes, having a brother named Mycroft, and so forth) or *exemplified* (initially being a creation of Conan Doyle, appearing in such and such novellas, etc.) by the fictional objects. The abstract character of resources is both upheld and denied in URI standards, especially in the RFC 2396, for instance. Far from being anecdotal, this betokens how difficult it is to overcome the received conception of objects as physical entities. It is directly from such a conception that seemingly stems the need to sunder resources into "information" and "non-information" ones, from UDIs to HTTP-Range 14 (though not REST!).

If we want to get rid of those paradoxes, we'll have to forge a conception of objects at least as subtle as what is hinted at with the concept of a resource.

3.2. Referentialization Instead of Reference

Up to now, the distinctions drawn from the architecture of the Web have been discussed in relation to the positions of three philosophers: Russell, Wittgenstein, and Kripke. According to this view, to understand the architecture of the Web one need only choose one position among the three. The problem with this view is that it deprives Web architecture of its autochthonous technical background, distinctions, and mediators; in other words, of its specificity being thus reduced to no more than an "intermediary" in Latour's sense, in other words the mere projection of conceptions drawn from the philosophy of language.

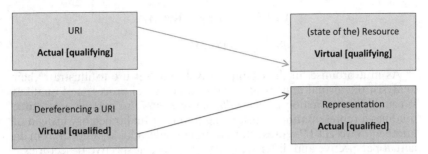

FIGURE 3.1. Referentialization of the Web.

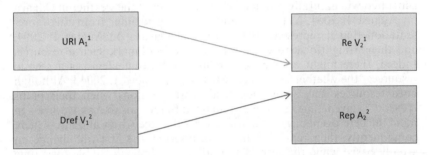

FIGURE 3.2. A process view on dereferentialization.

The work of Pierre Livet and Frédéric Nef (2009) on "social beings" (more aptly described as an attempt to "ontologize" Wittgenstein's concept of a rule) proved helpful in making sense of the architecture of the Web itself. Their "ontology of process" allows us to articulate resources and their representations according to their respective *virtual* and *actual* dimensions (the abstract resource having to do with the virtual, and its concrete representation with the actual—meaning that digital beings belong to the actual, the effective; in other words: *not* to the virtual).

In a nutshell, this analysis leads to a model where two processes converge, a qualifying process and a qualified one, as depicted in figure 3.1.

Translated into Livet and Nef's formalism,[4] the model takes the following form (depicted in figure 3.2):

[4] The arrow indicates a coupling, typical of processes, between the actual and the virtual; the characters between the parentheses refer to the aspect that is being replaced; the part in italics, to the aspect that will replace the latter; in bold, the process that is qualified; the superscript the initial (1) or final (2) aspect of a process; the subscript, whether it's process number 1 or 2; V stands for virtual, A for actual.

$$\text{URI } A_1{}^1 \rightarrow (Re \ V_2{}^1) \ \textbf{Rep } \textbf{A}_1{}^2$$

$$Rep \ A_2{}^2 \rightarrow \text{Dref } V_1{}^2$$

As in a promise, in the example Livet and Nef use to illustrate their analysis, the virtual dimension *weighs* on an actual process and qualifies it (qualifying an action *as the fulfilling of a promise*, just as a resource qualifies a representation as belonging to one virtual trajectory instead of another). Two HTTP representations might indeed look identical from an actual perspective and differ with regard to their respective trajectories.

Take this canonical example of Web architecture (Jacobs and Walsh 2004): a daily report of the weather in Oaxaca, Mexico. If one dereferences the URI that identifies this resource on August 1, 2004, the report generated will naturally provide information about the weather in Oaxaca on August 1, 2004. Yet one would be wrong to infer from this representation that it represents the weather of Oaxaca on August 1, 2004, and then straightforwardly use this URI to bookmark such a resource. Indeed, there might even be an archived URI that identifies a second resource: "the weather in Oaxaca, Mexico, on August 1, 2004." Although both resources may spawn identical representations when their paths cross at a given time, they will remain different insofar as we keep in mind the heterogeneous virtual trajectories that they draw and into which, consequently, their representations are inserted: the first will change on a daily basis, while the second typically *should* remain stable over time (on the Web, remaining stable is *never* a given; it always costs a lot).

Instead of the traditional notion of reference, the gap between words and the world, words and things, signs and objects, we're led to understand the basic relation between URIs and resources and their representations as a technical (and editorial) relationship of referentialization as displayed above, a relation where various mediators (servers, browsers, URIs, algorithms, standards, and so on) play a role that is simply impossible to conceive from the traditional perspective of the philosophy of language.

It belongs to a philosophy of (the architecture of) the Web to measure the discrepancy between itself and the tradition. The very practice of "philosophical engineering," a phrase coined by Tim Berners-Lee (2003) to designate the activity of Web architects, produces new distinctions and new entities without necessarily keeping to them, sometimes even burying their specificity behind received philosophical positions, whereas to remain faithful to philosophical engineering, one has *to go so far as to critically examine the philosophies under which it is buried by the actors themselves.* To summarize, philosophy is no longer treated as the *explanans* (the relation of reference that is studied by the philosophy of language) with regard to an *explanandum* (referentialization as a working, though largely modified, relation of reference; i.e., full of mediators, standards, shadows, and so forth).

3.3. The Object as a Rule

Philosophical engineering can be understood as the result of an *ontogonic* activity through which *new beings* emerge. At the center of Web architecture, the very notion of a resource explains the distinctions drawn in philosophical engineering by adding a level of indirection between the identifier and the tangible, actual, representations accessible on the Web. Instead of a dualistic picture, torn apart between identifiers understood as tags attached to ready-made, extensional objects, what is drawn in REST is a completely renewed picture of objects conceived as *rules of individuation*, reminiscent of Roy Fielding's definition of the resource as being "the semantics of what an author intends to identify" (Fielding and Taylor 2002, 135). The very act of positing an object means becoming engaged to remain faithful to that same object by *regularly* (in both the *normative* and the *temporal* senses of the word) serving adequate content *about* it. In other words, the main innovation of Web architecture, surprising as it may sound, is not to be found in any concrete mediator, be it the server or the browser, the URI or the link, HTTP as such, and so on. It is an *ontological innovation* around which everything else revolves and whose mode of existence in turn can only be felt through a coordinated choreography of mediators.

Assimilating objects to rules points to various philosophers, including Kant (see de Coorebyter 1985), for whom objects were constructed with rules—the concepts of the understanding, a list of which is given at the end of the Transcendental Deduction. In 1910, Ernst Cassirer (Cassirer 1977) extended the idea that *concepts* were rules through his notion of "serial objects," inspired by the progress of the modern logic of his time, in particular the privilege given to a functionalist symbolism over substances (here, the similarity to the quote above from Fielding is striking). Husserl (1989) made similar remarks at exactly the same time (1910) (see Gérard 2005). In 1928, Carnap (2001), in his *Aufbau*, completed this rethinking of objects and the shift from Kantianism to neo-Kantianism, the object looking more and more like a *function* insofar as it became impossible to distinguish it from a concept. Yet, despite the "construction" metaphor so paramount in Carnap's book, except for Cassirer and the late Husserl these thinkers neglected the importance of technics—a theoretical stance that can no longer be ours.

Here's the paradox: once again, the most autochthonous (or "heterogeneous," as Bernard Stiegler would put it in his trademark Husserlian fashion) components of the Web that emerged out of philosophical engineering pave the way for a renewed understanding of objectification, and thus of ontology itself. We need to understand what kind of referents there are on the Web. Although this is made possible through a careful analysis of its architecture, our investigations also lead to another paradox, since objects are no longer *physical objects*. Actually, they even look very much like *senses* or *meanings*. Yet we're not ready to adopt a neo-Fregean framework and abandon the focus on reference that seems to characterize the Web. Rather, it's time to think afresh what referents are. In this respect, an examination of the best system of reference humankind has ever known so far is a mandatory step.

David Kaplan's famous analysis of deictic expressions will help us to account for this little conundrum.[5] According to Kaplan, the meaning of a deictic expression is akin to a function, which doesn't vary. Kaplan distinguishes between the *content* of the indexical expression and its *character*. The character is a function or a "rule," the linguistic meaning of an expression that associates contexts as input and contents as outputs. Under such an analysis, an indexical like "I" may be analyzed this way:

	"I"
Character	The singularizing rule of being the enunciator of the expression "I"
Content	The object itself
Extension	The object itself

Kaplan's question consisted in determining whether when using deictic expressions they are being directly referred to or not. Despite his insistence on the importance of rules, objects, on this view, remain untouched. Therefore, we have to move one step further. The possibility is given by Brian Cantwell Smith's account of objects. Smith compares the aforementioned situation, with deictic expressions and rules, to the attraction exercised by a magnet. Just as with the rule of the deictic, the attraction remains constant and systematic (governed by laws, not unlike meaning itself). By contrast, the objects that fall within its attraction, its "referents," vary alongside the use of the magnet (the decisive factor being *use* more than *token*, says Smith). He contends that in both cases "the governing laws (regularity, habits) is an abstract but constant universal that maps particular occurrences—events, essentially—onto other particular occurrences or events, in a systematic way.... The crucial point of similarity, which is also the most difficult to say, has to do with the fact that the particularity of the result, referent of collected item, *spreads out through space and time*, in a kind of continuous egocentric (differential) way, until it captures the first entity that relates to the source or originating particular events in the mandated fashion" (Smith 1998, 170). Such individual entities remain beyond the grasp of physics; they are presupposed but not given. Whence this radical conclusion: "In sum, being an individual object is also not, in and of itself, an effective or even salient physical property. No physical attribute holds of an individual, for

---

[5] Kaplan's texts on direct reference are available in Davidson 2007.

example, except in virtue of its physical composition. If 'to be is to be the value of a bound variable,' physics will be of no help in easing the existential angst of any ordinary individuals. *For there are no physical objects*" (Smith 1998, 178; see also Livet and Nef 2009, 207).

Objects are no more given *hic et nunc* than resources are. Even more so, representations share with physical particularity the ability to spread out through space and time once they are related either to an object or to a resource accounting for their virtual trajectories. Hence, the Quinean motto that "to be is to be the value of a bound variable" (Quine 1980) may handily be turned into "to be an object is to be an individuating function" (or rule). Throughout innumerable encounters with representations, the object forever remains absent (at least *in toto*) in the background. Again, not very far from the way REST accounts for resources as intangible shadows and also reminiscent of the philosophical meaning of objects. "Objective existence" has been contrasted since the seventeenth century with "formal existence," what we would now call *concrete* existence, a trait Whitehead elaborated upon in the twentieth century, especially through his rejection of the principle of "simple location,"[6] according to which objects have a simple spatial and temporal location *hic et nunc*—a clear forerunner of our analysis if Web architecture is indeed an ontology.[7]

3.4. Frailty, Thy Name Is Resource

To impart a little more realism to this definition, one could refer to the work of Etienne Souriau, whose entire project, according to David Lapoujade, was to "save from nothingness the most frail and evanescent forms of existence" (Lapoujade 2011). We are balancing between two different worlds, to borrow a distinction made by Antoine Hennion (2007, 362): an "externalized world," shared, agreed upon, comprising autonomous entities, and an "internalized world," where nothing receives fixed properties or identity, where objects are constituted "by actively participating in constitutive operation." I would like to hint at a diplomatic mutual understanding of Wittgenstein's philosophy and Actor Network Theory by drawing a parallel between "attachments" as described by Hennion (2004) and "rules." Rules are typically in between objects and subjects, action

---

[6] See the analysis of Debaise (2006). Whitehead's discussion of this principle can be found in his *Science and the Modern World.*

[7] One should immediately add that the very word "ontology" appeared at the beginning of the seventeenth century, following the work of the late scholastic Francisco Suarez, in the works of Jacobus Lorhardus and Rudolph Goclenius. Contrasting with the tradition of the Aristotelian metaphysics, "ontology" was newly conceived as a "theory of objects" in the wake of Duns Scotus' concept of the univocity of being. My own work on the Web builds upon this understanding of the word "ontology" (*mutatis mutandis*).

and passion, freedom and determinism, and so forth. The parallel becomes all the more obvious when one realizes that *a rule is typically what makes someone do something*, in a very Latourian way of redefining agency, neither causally nor by sheer force. If resources are shadows, as much agent as patient of individuation, they nevertheless have their own agency, as befits objects—*and rules!* Nor are resources mere antifetishistic projections: they display a clear resistance by demanding *regular* coordination *of* their representations as well as *to* their representations—upon which they also depend. Not a small feat at all in the context of online publication.

It is easy to miss the resource and its *tentative ontology*, an ontology to be *achieved* rather than simply recorded. Forget good practices; substitute tangible mediators themselves to the *reason* why they are coordinated; or simply shift back to the world of extensional objects, "post-ontological objects" according to Smith (1998, 131)—and *voilà!* Even now discussions are going on inside the W3C to get rid of resources (see Summers 2013). Supposedly, Web developers have nothing to do with them—*in practice*; just as theoreticians stubbornly ignored them—*in theory*. To save an object from nothingness out of respect is a weird defense against a traditional ontological backdrop. Once we've moved to a completely different picture, call it the "successor metaphysics" as Smith does (1998, 87) or tentative ontologies as does Hennion, and the paradox disappears. All one needs is *to dispose oneself to be played by the rules.*

4. Conclusion: Toward Ontological Politics

There remains the task of determining how to agree on the components of a shared world. The solution doesn't come from a single person. Indeed, the idea to use Wikipedia to perform the function of "sorting propositions," which Latour identified in his *Politics of Nature*, dates back to 2006, with the birth of DBpedia, the central data repository of the entire Web of Data (as represented on the Linked Data Cloud).[8] This solution of using an encyclopedia in order to come to an agreement about the ontological furniture of the world is exactly what Ruth Barcan Marcus had in mind with her idea of a "dictionary," though this part of the philosophical problem was of no interest to her, as only the result (objects to be tagged) interested her, not the way we get there—unlike Semantic Web practitioners.

The political dimension of this ontological achievement should not be lost on us. Indeed, to produce an ontological device on a global scale

[8] Cf. http://lod-cloud.net/. See Auer et al. 2007.

cannot remain outside politics, because objects thereby acquire a political dimension, as pointed out by Noortje Marres:

> Like other forms of politics, we then say, the politics of objects is best approached as a performative politics. For no entity, whether human or non-human, institution or things, it suffices to posit on theoretical grounds that they "have" political capacities. For all entities, agential capacities depend at least in part on how these entities are equipped—on the configuration of an assemblage of entities that enable the explication of their normative capacities. This is why, somewhat paradoxically, in order to grasp the politics of objects, we must pay attention not just to objects, but also to the technologies and settings which enable them to operate. We must investigate how particular devices make possible the investment of things with political capacities. (Marres 2012, 104–5)

The Web, from the point of view of its architecture, certainly serves as a prime example of a device that "makes possible the investment of things with political capacities." Still, a lot remains to be done in order to improve the way objects and people are represented—in the political sense of the word; how the pluriverse, the many ways things both individuate themselves and are individuated, finds its adequate expression and related controversies made more visible. The latter could be achieved by making more "epistemic data" available on DBpedia regarding the state of the objectification on Wikipedia: the nature of the sources used to draw the portrait of an object (those who produce the primary sources quotable on Wikipedia in fact determine the nature of the *"porte-paroles"*—"experiments," "facts," "numbers," and so forth; others can mobilize to single out an object), the history of the articles, the discussions and controversies that they sometimes generate, and so forth.[9] As of now, these data are lost on us on DBpedia, as if no epistemic properties were involved in discriminating objects. We saw precisely the contrary.

The current philosophy of the Web, its architecture, appears not only as prosthesis for objectification. As is made obvious in the parallel with Brian Cantwell Smith's philosophy of objects, it is possible to think about it as a genuine philosophical position, despite its not being written in a book. We're no longer projecting philosophical concepts; rather, we came to recognize how the distributed agencies—rather than the actions—of "objects" (shadowy resources), "subjects" (publishers, Web architects, and so on), and mediators (standards, servers, protocols, languages, etc.) as much hold the Web as they hold thanks to the Web. Then, by extending

---

[9] The author is the co-initiator of the French version of DBPedia, available online at http://dbpedia.org and as part of the SemanticPedia project (http://www.semanticpedia.org/), a platform for the publication of the Wikimedia projects in French in conformity with Linked Data guidelines. This project presents an upgraded version of DBpedia, where the page history of each article is semanticized, and aims at keeping track of discussions as well both encouraging and fostering controversy analysis on Wikipedia.

the architecture of the Web, its operative philosophy or ontology, to Wikipedia (and DBpedia) understood as the institution where resources undergo various trials (what Joëlle Zask [2004] calls a process of "inter-objectification" in reference to John Dewey), we see how the Web performs something that would no longer qualify simply as an ontology (focusing on the nature objects *qua* objects) but as both a *political ontology* and an *ontological democracy* where everyone participates in determining and "making happen" what there is.

References

Auer, S., et al. 2007. "DBpedia: A Nucleus for a Web of Open Data." In *The Semantic Web: Lecture Notes in Computer Science*, edited by K. Aberer et al., 722–35. Berlin: Springer. Available at http://link.springer.com/chapter/10.1007/978-3-540-76298-0_52 (accessed June 9, 2013).

Berners-Lee, T. 1994. "RFC 1630—Universal Resource Identifiers in WWW: A Unifying Syntax for the Expression of Names and Addresses of Objects on the Network as Used in the World-Wide Web." Available at http://tools.ietf.org/html/rfc1630 (accessed July 1, 2009).

———. 2002. "ISSUE-14: What Is the Range of the HTTP Dereference Function?" Technical Architecture Group Tracker. *w3.org*. Available at http://www.w3.org/2001/tag/group/track/issues/14 (accessed May 12, 2012).

———. 2003. "Re: New Issue—Meaning of URIs in RDF Documents?" Public Discussion List of the W3C Technical Architecture Group. *w3.org*. Available at http://lists.w3.org/Archives/Public/www-tag/2003Jul/0158.html (accessed December 30, 2011).

Berners-Lee, T., R. T. Fielding, and L. Masinter. 2005. "RFC 3986—Uniform Resource Identifier (URI): Generic Syntax." Available at http://tools.ietf.org/html/rfc3986 (accessed July 1, 2009).

Berners-Lee, T., J. F. Groff, and R. Cailliau. 1992. "Universal Document Identifiers on the Network." Draft, CERN.

Berners-Lee, T., L. Masinter, and M. McCahill. 1994. "RFC 1738—Uniform Resource Locators (URL)." Available at http://tools.ietf.org/html/rfc1738 (accessed July 1, 2009).

Berners-Lee, T., et al. 1998. "RFC 2396—Uniform Resource Identifiers (URI): Generic Syntax." Available at http://tools.ietf.org/html/rfc2396 (accessed July 1, 2009).

Carnap, R. 2001. *La construction logique du monde*. Paris: Librairie Philosophique Vrin.

Cassirer, E. 1977. *Substance et fonction: Eléments pour une théorie du concept*. Paris: Les Éditions de Minuit.

Clark, K. G., 2002. "Identity Crisis." *XML.com*. Available at http://www.xml.com/pub/a/2002/09/11/deviant.html (accessed February 7, 2011).

————. 2003a. "Social Meaning and the Cult of Tim." *XML.com*. Available at http://www.xml.com/pub/a/2003/07/23/deviant.html (accessed January 21, 2011).

————. 2003b. "The Social Meaning of RDF." *XML.com*. Available at http://www.xml.com/pub/a/2003/03/05/social.html (accessed January 21, 2011).

de Coorebyter, V. 1985. "La doctrine kantienne du concept en 1781." *Revue Philosophique de Louvain* 83, no. 57:24–53.

Cussins, A. 2001. "Norms, Networks, and Trails: Relations Between Different Topologies of Activity, Kinds of Normativity, and the New Weird Metaphysics of Actor Network Theory: And Some Cautions About the Contents of the Ethnographer's Toolkit." In *Keele Conference on Actor Network Theory*. Available at http://www.haecceia.com/FILES/ANT_v_Trails_jan_01.htm (accessed August 22, 2012).

Davidson, M. 2007. *On Sense and Direct Reference: Readings in the Philosophy of Language*. New York: McGraw-Hill.

Debaise, D. 2006. *Un empirisme spéculatif: Lecture de Procès et Réalité de Whitehead*. Paris: Librairie Philosophique Vrin.

Delany, P., and G. P. Landow. 1991. *Hypermedia and Literary Studies*. Cambridge, Mass.: MIT Press.

Fielding, R. T. 2000. "Architectural Styles and the Design of Network-Based Software Architectures." Ph.D. dissertation, University of California, Irvine. Available at http://www.ics.uci.edu/%7Efielding/pubs/dissertation/fielding_dissertation.pdf (accessed January 13, 2009).

————. 2005. "[httpRange-14] Resolved." Available at http://lists.w3.org/Archives/Public/www-tag/2005Jun/0039.html (accessed December 24, 2012).

————. 2006. "Re: RFC for REST?" Available at http://permalink.gmane.org/gmane.comp.web.services.rest/3995 (accessed December 19, 2012).

————. 2008. "REST APIs Must Be Hypertext-Driven." *Untangled*. Available at http://roy.gbiv.com/untangled/2008/rest-apis-must-be-hypertext-driven (accessed January 21, 2011).

Fielding, R. T., and R. N. Taylor. 2002. "Principled Design of the Modern Web Architecture." *ACM Transactions on Internet Technology (TOIT)* 2, no. 2:115–50.

Fielding, R. T., et al., eds. 1997. "RFC 2068 Hypertext Transfer Protocol—HTTP/1.1." Available at http://tools.ietf.org/rfc/rfc2068.txt (accessed May 12, 2012).

Fielding, R. T., et al., eds. 1999. "RFC 2616 Hypertext Transfer Protocol—HTTP/1.1." Available at http://tools.ietf.org/rfc/rfc2616.txt (accessed May 12, 2012).

Gérard, V. 2005. "Husserl et la phénoménologie de la choséité." In *La connaissance des choses: Définition, description, classification*, edited by G. Samama, 139–49. Paris: Ellipses Marketing.

Halpin, H., 2009. "Sense and Reference on the Web." Ph.D. thesis, Institute for Communicating and Collaborative Systems, School of Informatics, University of Edinburgh. Available at http://www.ibiblio.org/hhalpin/homepage/thesis/ (accessed November 11, 2009).

Hennion, A. 2004. "Une sociologie des attachements." *Sociétés* 85, no. 3:9–24. Available at http://www.cairn.info.domino-ip2.univ-paris1.fr/revue-societes-2004-3-page-9.htm (accessed July 1, 2012).

———. 2007. *La passion musicale: Une sociologie de la médiation.* Édition revue et corrigée. Paris: Editions Métailié.

Hennion, A., and B. Latour. 1993. Object d'art, object de science. Note sur les limites de l'anti fétishisme *Sociologie de l'art* 6:7–24.

Husserl, E. 1989. *Chose et espace: Leçons de 1907.* Paris: Presses Universitaires de France.

Jacobs, I., and N. Walsh. 2004. *Architecture of the World Wide Web, Volume One (W3C Recommendation 15* December 2004). Available at http://www.w3.org/TR/webarch/#formats (accessed February 1, 2009).

Janicaud, D. 2005. *Heidegger en France*, vol. 2. Paris: Hachette.

Koepsell, D. R. 2003. *The Ontology of Cyberspace: Philosophy, Law, and the Future of Intellectual Property.* New edition. Chicago: Open Court.

Kripke, S. 1980. *Naming and Necessity.* Cambridge: Harvard University Press.

Landow, G. P. 1991. *Hypertext: Convergence of Contemporary Critical Theory and Technology.* Baltimore, Md.: Johns Hopkins University Press.

———. 1994. *(Hyper/Text/Theory).* Baltimore, Md.: Johns Hopkins University Press.

———. 1997. *Hypertext 2.0: Convergence of Contemporary Critical Theory and Technology.* Second revised edition. Baltimore, Md.: Johns Hopkins University Press.

———. 2006. *Hypertext 3.0: Critical Theory and New Media in a Global Era.* Third revised edition. Baltimore, Md.: Johns Hopkins University Press.

Lapoujade, D. 2011. "Etienne Souriau, une philosophie des existences moindres." In *Anthologies de la Possession*, edited by Didier Debaise, 167–96. Paris: Presses du Réel.

Latour, B. 2005. *La science en action: Introduction à la sociologie des sciences.* Nouvelle édition. Paris: Editions La Découverte.

Lewis, R., ed. 2007. *Dereferencing HTTP URIs (Draft Tag Finding 31 May 2007).* Available at: http://www.w3.org/2001/tag/doc/httpRange-14/2007-05-31/HttpRange-14 (accessed January 21, 2011).

Livet, P., and F. Nef. 2009. *Les êtres sociaux: Processus et virtualité.* Paris: Hermann.

Marcus, R. B. 1995. *Modalities: Philosophical Essays.* New York: Oxford University Press.

Marres, N. 2012. *Material Participation: Technology, the Environment and Everyday Publics*. Basingstoke: Palgrave Macmillan.

Monnin, A. 2013. "Vers une Philosophie du Web: Le Web comme devenir-artefact de la philosophie (entre URIs, Tags, Ontologie(s) et Ressources)." Ph.D. thesis, Université Paris 1 Panthéon-Sorbonne.

Putnam, H. 2005. *Ethics Without Ontology*. New edition. Cambridge, Mass.: Harvard University Press.

Putz, S. 1993. "Design and Implementation of the System 33 Document Service." Technical Report ISTL-NLTT-93-07-01, Xerox Palo Alto Research Center.

Putz, S. B., M. D. Weiser, and A. J. Demers. 1993. United States Patent: 5210824. Encoding-format-desensitized methods and means for interchanging electronic document as appearances. Available at http://patft.uspto.gov/netacgi/nph-Parser?Sect2=PTO1&Sect2=HITOFF&p=1&u=/netahtml/PTO/search-bool.html&r=1&f=G&l=50&d=PALL&RefSrch=yes&Query=PN/5210824 (accessed October 2, 2012).

Quine, W. V. 1980. *From a Logical Point of View: Nine Logico-Philosophical Essays*. Second revised edition. Cambridge, Mass.: Harvard University Press.

Shadbolt, N. 2007. "Philosophical Engineering." In *Words and Intelligence II: Essays in Honor of Yorick Wilks*, edited by Khurshid Ahmad, Christopher Brewster, and Mark Stevenson, 195–207. Berlin: Springer.

Smith, B. C. 1998. *On the Origin of Objects*. Cambridge, Mass.: MIT Press.

Sollins, K., and L. Masinter. 1994. "RFC 1737—Functional Requirements for Uniform Resource Names." Available at http://tools.ietf.org/html/rfc1737 (accessed July 1, 2009).

Summers, E. 2013. "Linking Things on the Web: A Pragmatic Examination of Linked Data for Libraries, Archives and Museums." *CoRR*, abs/1302.4591. Available at http://dblp.uni-trier.de/rec/bibtex/journals/corr/abs-1302-4591 (accessed June 9, 2013).

Thévenot, L. 1984. "Rules and Implement: Investment in Forms." *Social Science Information* 23, no. 1:1–45.

Vuillemin, J. 2009. *What Are Philosophical Systems?* Cambridge: Cambridge University Press.

Zalta, E. N. 2003. "Referring to Fictional Characters." *Dialectica* 57, no. 2:243–54.

Zask, J. 2004. "L'enquête sociale comme inter-objectivation." In *La croyance et l'enquête: Aux sources du pragmatisme*, edited by Bruno Karsenti and Louis Quéré, 141–63. Paris: Presses de l'EHESS.

CHAPTER 4

WHAT IS A DIGITAL OBJECT?

YUK HUI

In this chapter I attempt to outline what I call digital objects, showing that a philosophical investigation is necessary by revisiting the history of philosophy and proposing that it is possible to develop a philosophy of digital objects. I consider first the question of the digital, then the question of objects, and finally the question of the digital again. What I call digital objects are simply objects on the Web, such as YouTube videos, Facebook profiles, Flickr images, and so forth, that are composed of data and formalized by schemes or ontologies that one can generalize as metadata. These objects pervade our everyday life online, and it is in fact very difficult for us to separate what is online and offline anymore, as indicated decades ago by the action of "jacking into cyberspace."[1]

It is not only that we become addicted to different trendy gadgets, but also that they constitute a ubiquitous milieu from which we cannot escape. Digital objects are not simply bits and bytes, as proposed in the digital physics or digital ontology in the works of Edward Fredkin and Stephen Wolfram. Digital ontology consists of two main concepts: first, that bits are the atomic representation of the state of information; and second, that the temporal state of evolution is a digital information process (Floridi 2009). The second point embodies a long historical debate between humanism and cybernetics. Nevertheless, both concepts ignore the fact that we are interacting with digital objects: they are actually objects that we drag, we delete, we modify, and so on. The Web is acting both as an interface between users and digital objects and as a world in which these digital objects conceal and reveal—in both physical and metaphysical terms. But I am not suggesting here that the previous propositions about the digital are simply wrong; to use an analogy, we now know that the world consist of atoms, but to think only in terms of atoms won't help us to explain the world. That is to say, such a digital philosophy is insufficient to help us reach an understanding of everyday life amid technological acceleration, not to mention a deeper reflection on existence.

---

[1] A phrase used frequently by William Gibson in *Neuromancer* (1984), which nicely describes the separation between two worlds that one tended to imagine in the 1980s and 1990s.

In this chapter, I propose first to move the investigation from the digital to objects, and continue from there. Then I want to contrast digital objects with past investigations into natural objects and technical objects, and finally I will extend the analysis to digital objects. First of all, I want to make a not-so-modest claim here that Western philosophy from Aristotle to Edmund Husserl concerns only natural objects, or more precisely how the objects appear or are shown to us. So first let us look at the question of natural objects. When speaking of natural objects, we don't mean objects given by nature, such as vegetables or animals. A natural object here refers to the category in which every object, whether natural or fabricated, is analysed in the same natural manner. This method proposes that an object can be understood by grasping its essence, which determines its particular appearance. This process of knowing, at first glance, already supposes the object itself and the object for knowledge. This leads to the development of a scientific knowledge that works towards an absolute certainty, which guarantees the correspondence between the thing itself and consciousness. In his *Categories* Aristotle proposes to understand being in terms of substance and accident. He says: "That which is called a substance most strictly primarily and most of all—is that which is neither said of a subject nor in a subject, e.g. the individual man or the individual horse" (Aristotle 1984, 2a13–2a18). Substance itself is the subject. Accidents are the predicates of the subject. Clearly, in his *Categories* Aristotle designates the subject-predicate pairing both as a grammatical structure and as a system of classification. The relation between language as classification and things as physical beings is already established.

Aristotle gives a more detailed, while somewhat divergent, account of substance in *Metaphysics* (book Z), where he says that the question "'what is being?' really amounts to 'what is substance?'" (Aristotle 1956, 168). He then proposes to understand the substance of the substratum. The substratum can be described in terms of sensible form and matter. Sensible form is concerned with "what kind of thing" something is, and matter concerns "what it is made of." Aristotle proposed to decide which of the three elements, form, matter, or the composite of form and matter, can be called substance. He rejected matter and the composite of matter with form, the first because it can be a predicate of the subject, the second because it is "posterior in nature and familiar to sense" (Aristotle 1956, 172). He finally decided that form is the sole understanding of substratum. Sensible forms raise the question of essence. There are two points we have to note here: first, the question of substantial form became a long-lasting philosophical question concerning the essence of things and their representation; and second, the distinction between subject and object did not come to be made until Descartes, and so the thing under contemplation is a subject but not an object. The concept of subject moving away from thing to the ego that contemplates it is

characteristic of a separate yet constant mediation between subject (consciousness) and substance (essence) (Rotenstreich 1974, 2).

The subject-substance question can be understood as the core of the
philosophical conceptuality of natural objects (Rotenstreich 1974, 1). We
can follow a long historical trajectory from Hume through Kant,
German idealism (including Fichte, Hegel, and so on), and later Husserl,
which one can call the phenomenological tradition. These philosophers
proposed different models for understanding the relation between subject
and substance, and it is obvious that one cannot generalize their thought,
since each of them requires considerable investigation. However, if there
is something one can say these philosophers have in common it is that
they all want to find out how the subject allows substance to manifest
itself as such, and how the subject takes a more and more active role (for
Hume, the subject is almost passive). As we cannot undertake a thorough
examination of the thought of each philosopher one by one here, I would
like to exemplify this tradition through Husserlian phenomenology,
since Husserl is the one who made "Back to the things themselves!" the
slogan of phenomenology. Husserlian phenomenology is known as
descriptive phenomenology. The very word "descriptive" clearly distinguishes Husserl from Hegel. For Husserl, phenomenology is a descriptive
process, which goes back and forth to depict the object through the
knowing consciousness, while for Hegel phenomenology is a speculative
process in which multiple stages of self-consciousness are attained
through dialectical movements and sublations. They are not totally separated, however, since Husserl's phenomenology is another investigation
into consciousness and is an attempt to provide the absolute foundation
of all science. From this perspective Husserl and Hegel share the same
ambition.[2]

Husserl's phenomenology rejected Kant's thing-in-itself (*das Ding an
sich*), which states that human beings can know only the phenomenon
of things; knowledge of thing-in-itself demands an intellectual intuition
which is absent in human beings (Kant 1996). Husserl denounced the
thing-in-itself as a mystery, and he proposed that we can actually know the
object through the movements of intentionality. Since Husserl starts as an
arithmetician, then becomes a philosopher of logic and consciousness, and
finally ends as a philosopher of culture, it is almost impossible to summarize a theory of the object in a way that captures his entire understanding.
But in a nutshell, Husserl regards everything as a possible intentional
object; for example, a number or an apple is an object. Husserl's project is
directed against what he calls naïve realism and relativism. An object for
Husserl is not what is given; rather, this given is constituted by a genesis of

---

[2] Husserl's connection with Hegel, in my view, can be made through Heidegger, especially through his understanding of Hegel's concept of "experience" and Husserl's notion of
"categorial intuition."

the senses. In order to relinquish naïve realism, the phenomenologist starts with *epoché*, meaning bracketing any presupposition and bias, which already constitutes the object as such. The bracketing process, to Husserl, is also a process of returning to an absolute Ego, which is free from any presupposition. In this sense the subject takes a much more active role. An intentional act then comes into being, directed from the subject to the object, and the reflection that this act effects constitutes a horizon on which the ideality of the object appears. This ideality is only possible through a process of ideation,[3] which reconstitutes the horizon.

The trajectory of the modern metaphysics of objects opens up several general directions for the investigation of objects. First, there is a wavering scepticism regarding the concept of substance. The transcendence of substance finds its location in God; in other words, substance and God are on the same plane, since they are beyond human experience. The risk involved in an absolute knowledge of the object easily leads to the destruction of the whole plane by bringing it down to the plane of immanence. This philosophical trajectory also accompanies the scientific spirit in working towards the discovery and reassurance of the power of scientific methods, which create an exclusive system of knowledge. Second, consciousness is the ultimate mystery, and no authority can describe for itself the ultimate truth for ever. These multiple models attempt to comprehend the mind, and they assign different mechanisms to it. This is important, since the mind is the same as the object of inquiry (even if it is much more complicated), and we can also pose the question of the thing-in-itself of the mind just as we may do for a fillet of steak or a cauliflower. In Hume, Kant, Hegel, and Husserl, consciousness is imbued with specific functions, which are also systemized as part of an *organon* of knowing (although none of them would admit the word "organon"). Third, the role of knowing falls totally on the mind. The other side of the coin is that objects are always objects of experience. The predicates of the objects are qualities that can be experienced, so all of the above-mentioned philosophers are eager to find the structure of consciousness that would allow it to know the object, but there is among them less investigation into the object's own existence, and how its existence conditions the process of knowing and being itself.

Technical Objects

Within the dialectics of substance and subject, there is no place for technical objects. Ignorance of these objects in philosophy has meant that it has failed to absorb the rapid development of technology and social

---

[3] Ideation here takes a Platonic sense, meaning how the ideality can be deduced through a mediative process such as recollection; for Husserl, it consists of different cognitive functions such as explication, negation, and so on, that seek the essence of the object.

change after the industrial revolution. The idea of the philosopher as a figure who stands outside as mere critic and defends the purity of thought and inquiry into human nature has been washed away in the flux of technological progress. It is possible to argue that most of the philosophers of phenomenology except Husserl came before the industrial revolution, so they dismissed technical objects. Yet technical objects are not necessarily complicated machines; a hammer or a knife is also a technical object. Indeed, Husserl the philosopher witnessed the rapid proliferation of machines after the industrial revolution but didn't bring them into his phenomenological theory.[4] A new philosophical attitude as well as a new philosophical system must be constituted in order to comprehend the changes that this process entailed. If ontology starts with the question of being, then there is a problem that the understanding of being is not on the right path if it does not take into account the nature of technology. And this is very clear if we follow Heidegger's proposition that the beginning of cybernetics is the end of metaphysics (Heidegger 2001). I will therefore propose two figures who may bring the concept of technical objects to light and prepare the ground for our investigation of digital objects: the French philosopher Gilbert Simondon (1924–1989) and the German philosopher Martin Heidegger (1889–1976). They may appear at first glance to be incompatible, because Simondon is an admirer of modern technology, while Heidegger is known as a philosopher who was opposed to it.

Simondon's 1958 doctoral thesis, later published as *On the Mode of Existence of Technical Objects* (1980), proposed what he calls a "mechanology." Mechanology investigates the existence of technical objects through their movement towards perfection. Simondon demonstrates their lineage from the origin of technology to the point where it provides an increasingly concrete object through the example of the evolution from diodes to Lee de Forest triodes. The diode is a device that controls the flow of current in a single direction. In its simplest form, within a vacuum tube, the cathode is heated and hence activated to release electrons. The anode is positively charged so that it attracts electrons from the cathode. When the voltage polarity is reversed, the anode is not heated, and thus cannot emit electrons. Hence there is no current. A triode places a grid between the anode and the cathode; a DC current can give a bias to the grid: if it is negative, it will repel some of the electrons back to the cathode and hence serve as an amplifier (see figure 4.1). Simondon proposes that the absolute beginning of the triode is not the diode but is to be found "in

---

[4] The absence of technical objects in Husserl is further elaborated by Bernard Stiegler in *Technics and Time*, vol. 1 (1998) and vol. 2 (2009). Stiegler shows that Husserl was able to talk about primary and secondary retention but not tertiary retention, which is one of the most important elements of technical objects.

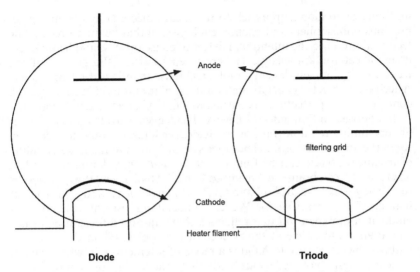

Diode   Triode

FIGURE 4.1. An indirect heated vacuum tube diode and triode.

the condition of irreversibility of the electrodes and the phenomenon of the transport of electric charges across the vacuum" (Simondon 1980, 36).

The diode or the triode is what Simondon calls a technical element, and the ensemble of these elements constitutes a technical individual. But one shouldn't simply understand it as a collection of components; a technical individual is a technical object that supports the functioning of its inner structure at the same time as it is able to adapt an external milieu to its functioning. This view differs from the views of some other theorists on technical compositions, such as Herbert A. Simon. Simon approaches technologies through systems and subsystems, and the interface that allows subsystems to communicate with each other (Haugeland 1993). Simondon goes deeper down to the modes of existence of technical objects and derives a theory of system from different "orders of granularities" ranging from technical elements, to individuals, to systems. What is the most intriguing and most interesting thing in Simondon's theory of systems is the idea of an "associated milieu" that provides a stabilizing function to restore the equilibrium of the system itself. For example, Simondon often spoke of the Guimbal turbine (named after the engineer who invented it), which, to solve the problem of loss of energy and overheating, uses oil to lubricate the engine and at the same time isolate it from water; it can then also integrate a river as the cooling agent of a turbine (Simondon 2005). The river here is the associated milieu for the technical system; it is part of the system but not a component of the machine. Simondon's approach to technical objects differs from that of previous philosophers and phenomenologists in that he didn't reduce the technical object to the intentional defect of consciousness and hence make it an object for knowledge. He proposed to study the genesis of the technical object itself, less in a biological sense than in a mechanical one. A technical object regains its materiality and attains a different degree of concreteness or perfection in contrast to what cybernetics term "control."

In contrast to Simondon, I believe, Heidegger provided a new way of understanding relations (although Heidegger himself would immediately reject the above proposal). Heidegger's contribution to the understanding of technical objects can be found in *Being and Time*, dating from 1927, where he talks about the "ready-to-hand." Heidegger (1967) proposes two categories: "ready-to-handness" (*Zuhandenheit*) and "present-at-handness" (*Vorhandenheit*). We can understand present-at-hand as a mode of comprehension that renders a thing an object for consciousness and attempts to arrive at the essence of that object (as in the case of a natural object). Ready-to-hand is a mode of interaction, in which we put aside the question of ideality and objectivity and let the object appear to us according to its functionalities. We see a similar impulse in Simondon and Heidegger here, characterized by a move from substance to external milieu, which allows the object to be defined.

The difference between them is that Heidegger bypassed the technical milieu and concentrated on the social milieu, and he interpreted the object's self-manifestation within its milieu in terms of human *Dasein*. For example, Heidegger illustrates the way we use a hammer: we don't really need to achieve an ideality of the hammer (as present-at-hand) before we use it; we just grasp it and use it to hit the nail into the place it is intended. This daily practical activity moves away from the concept of experience as a mere activity of consciousness, arguing that the previous understanding of objects which subsumed them under cognition ignores the world of both objects and *Dasein*. For instance, according to Heidegger, Husserl's concept of intentionality when properly understood is nothing but the awareness of being-in-the-world; that is to say, it is not a ray projected from the ego but a field from which the ego cannot escape (Heidegger 1988). Heidegger's approach towards technical objects was taken up by philosophers such as Maurice Merleau-Ponty and Hubert Dreyfus, and later by AI researchers as a challenge in the design of intelligence.

Digital Objects

Both investigations into natural objects and technical objects in the phenomenological tradition show us different directions in which objects could be studied. Digital objects are visible to us in different forms. We can treat them as natural objects. They demand the engagement of our consciousness to furnish concepts for their appearance and our experience with them. Following the phenomenology of Kant, Hegel, and Husserl, we can investigate the movement of reason and intentionality. The previous theories regarding natural objects still have their place. But if these kinds of investigation are still possible, are they sufficient to address the question of digital objects? What can we think about the "substance" of a digital object? Digital objects appear to human users as colourful and visible beings. At the level of programming they are text files; further down the operating system they are binary codes; finally, at the level of circuit boards they are nothing but signals generated by the values of voltage and the operation of logic gates. How, then, can we think about the voltage differences as being the substance of a digital object? Searching downward we may end up with the mediation of silicon and metal. And finally we could go into particles and fields. But this kind of reductionism doesn't tell us much about the world.

Following the Simondonian approach, we can produce a genesis of digital objects by studying the evolution of technical apparatus, for example, metadata schemes; with Heidegger, the objects constitute the milieu that we are living in, giving us a new interpretation of being-in-the-digital-milieu. But first of all we must grasp the specificity of digital objects and from there make these connections clear. I want to go back to the question of the digital again, and propose that one fails to see the whole landscape if one simply understands the digital as only a 0 and 1 binary code; rather, one should grasp the digital as a new technique to manage data in comparison with the analogue. The French philosopher Bernard Stiegler follows the French anthropologist Sylvain Auroux in proposing the idea of grammatization, which "designates more general the passage from temporal continuous to spatial discrete, a fundamental form of the exteriorization of flux to" tertiary retention.[5] Stiegler further classifies three discretenesses of grammatization, namely: literal, analogue, and digital. These levels of discreteness designate different systems of writing and reading, and, more important, the ways of exteriorization and the possibilities opened up thereby. Thinking in terms of exteriorization gives us a significant clue to move away from the analysis of natural and partially technical objects.

When we look at the term "data" we hardly recognize that its Latin root *datum* originally means "[a thing] given"; the French word for data, *donnée*, has this meaning as well. If data are the things given, what gives them? This is the question for both investigations into natural objects and technical objects: for natural objects, the given is closely related to sense data; among the theorists of technical objects mentioned above, Heidegger attempts to propose givenness as the condition of the appearance of the world that gives rise to a new interpretation of the relation between human beings and things. But we have to recognize that since 1946 the word

---

[5] "Grammatisation," at http://arsindustrialis.org/grammatisation

"data" has had an additional meaning: "transmittable and storable computer information."[6] This second sense of "data" suggests a reconsideration of the philosophy of objects, since the givenness can no longer be taken as sense data or a mode of being together of man and nature; instead, one has to recognize its material transformation. The significance of the new technique of data processing we now call the digital is not only that with computers we can process large amounts of data but also that by operating with data the system can establish connections and form a network of data that extends from platform to platform, database to database. The digital remains invisible without data, or traces of data. With the population of Web-based applications (further amplified by social networking), the production of data is increasing in a manner that one can hardly imagine. Let me quote Berkeley computer science professor Michael Franklin on the production of data by a single user, from which we can get a glimpse of the universe of data we are living with:

> Most tweets, for example, are created manually by people at keyboards or touchscreens, 140 characters at a time. Multiply that by the millions of active users and the result is indeed an impressive amount of information. The data driving the data analytics tsunami, on the other hand, is automatically generated. Every page view, ad impression, ad click, video view, etc. done by every user on the web generates thousands of bytes of log information. Add in the data automatically generated by the underlying infrastructure (CDNs, servers, gateways, etc.) and you can quickly find yourself dealing with petabytes of data.[7]

Users are producing tremendous amount of data, physical objects are becoming fact-based data, by digitization, RFID tags, and so on; fact-based data are becoming digital objects, meaning that data must be conceptualized as graspable entities by both the human mind and the computational mind. These two processes are what I call the *datafication of objects* and the *objectification of data*. The question in the engineering sense is, What is the best way to manage data? Transformed into a philosophical question, How concrete should the objects be? We can see that Web ontologies that present in the form of GML, SGML, HTML, and XML and more recently Web ontologies under the name Semantic Web are endeavours to create different levels of concreteness (Berners-Lee 2001), and networks in which each relation can be articulated and calculated. This evolution process is not linear at all; every progress is conditioned by the technical milieu. From GML to HTML we actually see a loss of concreteness: since HTML tends to be a lightweight language, it

---

[6] Online Etymology Dictionary, http://www.etymonline.com/
[7] Quoted by Ben Lorica, "Big Data and Real-Time Structured Data Analytics," www.radar.oreilly.com/2009/08/big-data-and-real-time-structured-data-analytics.html, accessed 14 December 2011.

tends to reduce objects to representations and use only hyperlinks as relations in the networks. From HTML to XML and Web ontologies, the objects are becoming more and more concrete, if by concrete here we mean that the concepts of the objects are more well defined and the relations between parts of the objects and between objects are more explicit—that is, no longer limited by hyperlinks but by parsing and comparing well-structured data.

Horizontally, we can see that as the associated milieu enlarges in terms of quantities through the progress from GML (for compatibility between programs within a machine) to ontologies (across the Internet between machines), it involves more and more objects, machines, and users to maintain its functionality and stability. We can also think of the associated milieu as a measurement of interoperability and compactability here. *Vertically*, digital objects are always in a process by which they become more concrete and individualized. Concretization for Simondon also means increasing levels of abstraction; in ontologies, we find that there is ambiguity between a computer programme and a text file. HTML is simply a formatted text file full of data, but RDF (resource definition framework) defines complicated documents with programming and logical capacities. Ontologies in the RDF or OWL (Web ontology language) format become similar to an object in OOP (object-oriented programming), which has three important properties, namely, abstraction, encapsulation, and inheritance (a class can be overridden to generate new classes, which inherit certain properties and functions of the parent class), and we can identify all of these in an ontology.

The function of structured data is to produce formal relations between each relation in RDF no matter how arbitrary these relations are. For example, even "difference" can become a formal relation for comparison. As we have seen, a digital object is also a natural object, which possesses different qualities. These qualities are represented in the form of data and metadata. The relation between data and metadata has to be further distinguished. By definition metadata are data about data. That is to say, they are a description of something else. But this description can extend infinitely and may end up as circular. It is also this infinite extension of "data of data" that constitutes a different network. Being also computational objects, digital objects are subsumed under calculation. The affectivity and sensibility of the objects are calculable. The metadata of a digital object can grow in time if the database assigns more attributes to it. But at least its relation to other digital objects will increase, even though it remains the same. When there are more digital objects, there are more relations, hence the networks either become larger or new networks are actualized. An object is meaningful only within a network; for example, a Facebook invitation is meaningless if there is not a network that is mediated by the data of the users. The multiple networks, which are connected by protocols and standards, constitute what I call a digital milieu. (See figure 4.2.)

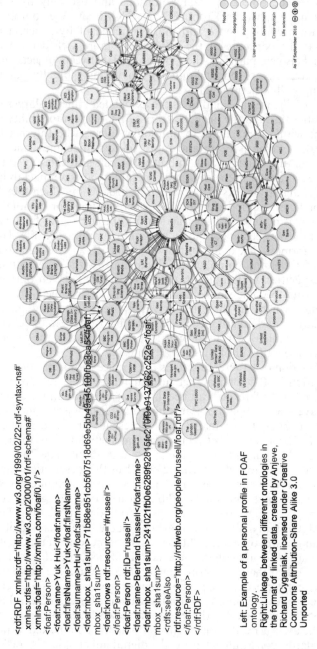

```
<rdf:RDF xmlns:rdf="http://www.w3.org/1999/02/22-rdf-syntax-ns#"
xmlns:rdfs="http://www.w3.org/2000/01/rdf-schema#"
xmlns:foaf="http://xmlns.com/foaf/0.1/">
<foaf:Person>
<foaf:name>Yuk Hui</foaf:name>
<foaf:firstName>Yuk</foaf:firstName>
<foaf:surname>Hui</foaf:surname>
<foaf:mbox_sha1sum>71b88e951cb5f07518d69e5bb49545f90fb63ca5</foaf:
mbox_sha1sum>
<foaf:knows rdf:resource= "#russell"/>
</foaf:Person>
<foaf:Person rdf:ID="russell">
<foaf:name>Bertrand Russell</foaf:name>
<foaf:mbox_sha1sum>241021fb0e6289f92815fc21bf0e9137262c252e</foaf:
mbox_sha1sum>
<rdfs:seeAlso
rdf:resource="http://rdfweb.org/people/brussell/foaf.rdf"/>
</foaf:Person>
</rdf:RDF>
```

Left: Example of a personal profile in FOAF
ontology.
Right:Linkage between different ontologies in
the format of linked data, created by Anjeve,
Richard Cyganiak, licensed under Creative
Commons Attribution-Share Alike 3.0
Unported

FIGURE 4.2. Objects are data, data are sources of networks. A personal profile in the format of FOAF (friend of a friend) ontology, and ontologies in the format of linked data connected to form networks across multiple domains.

Not Yet a Conclusion

Let me provide a brief summary of what we've been discussing. Digital objects appear in three phases, which are interdependent of each other but cannot be reduced or generalized into oneness: *objects*, *data*, and *networks*. If the investigation of natural objects is concerned with the dialectics of subject and substance, and the investigation of technical objects is concerned with the relationality between the object and the milieu, then the investigation of digital objects must obtain a new direction by pushing these two investigations further. This does not mean that the previous investigations lose their significance; it simply indicates that the question of substance is no longer at issue, since it is not only undemonstrable, as Hume showed, but also unthinkable. The investigation of digital objects must find a new relation between object and mind. Furthermore, the relationality within technical objects and their relationality to the world are not independent of each other. Technical objects are not only symbols as they appear in the world, nor are they simply tools for use; their internal relations are materialized and codified, which in turn conditions the opening of the world. This opens up many different inquiries towards a philosophy of digital objects, and here I want to specify two of them.

The first concerns what Bernard Stiegler calls tertiary retention, Andy Clark and David Chalmers's idea of the extended mind (Clark and Chalmers 1998), John Haugeland's embedded mind (Haugeland 1993), and Fred Dretske's externalism (Dretske 2004). More specifically, we are talking here about digital objects as externalized memories that condition our retrieval of the past and anticipation of the future. If traditional phenomenology, especially that of Hegel and Husserl, gives the subject an active role of knowing and experience, then the reconsideration will bring the subject back to its passive mode and give us a higher position to digital objects. This doesn't deprive the subject of its role of cognition, it attempts to understand the condition of cognition. I single out Stiegler's theory because, by comparison with the others, it poses the political question of externalities. Philosophers such as Clark, Chalmers, Dretske, and Haugeland come from an AI perspective, especially one that is haunted by Heideggerian AI. As we briefly saw in some of the propositions of Heidegger in the section above on technical objects, Heideggerian AI argues that the good old-fashioned AI went totally wrong because it understood cognition through a Cartesian approach, seeing the mind as the source of the production of all meaning and its essential part as responsible for creating representations of the world. Instead, Heideggerian AI holds the view that the world itself is the source of meaning that conditions human actions, and the way human beings interact with the world is not necessarily mediated by representations, for example, the use of a hammer as described by Heidegger. Identifying with a Heideggerian or pseudo-Heideggerian spirit, Clark proposes the idea of scaffolding to describe how the mind operates beyond the skull and the skin; Dretske proposes that what is important for a tool is not that something is represented by it but rather "how it represents the world" (Dretske 2004, 397); Haugeland seizes the concept of "affordance" from J. J. Gibson's ecology of perception, which suggests looking at meaning given to us by the environment, rather than derived from our human speculation (Haugeland 1993).

By comparison, Stiegler's theory of externalities is much more informed by Husserl rather than by Heidegger, and among others especially by the anthropologist André Leroi-Gourhan, who explores the physiological development of human beings in relation to the use of tools. In order to look into Stiegler's tertiary retentions, we must go back to Husserl's system of time consciousness (*Zeitbewusstsein*). To explain Husserlian time consciousness in a nutshell here, let's imagine that we are listening to a melody; we are experiencing a flux of consciousness, which is the passing of the "nows." The "now" that is retained immediately in my mind is what Husserl calls primary retention, the melody that I can recall tomorrow is called secondary retention; these retentions condition protentions as well, which also means anticipations and projections. Tertiary retention supplements the finitude of the first two kinds of retention with an infinite repertoire of memories, made possible by digitization. But on the other hand, the tertiary retention is also the source of the primary retention, and the support of the secondary retention, which is also the source of protention. The causality of intentionality is taking on a new configuration. Hence Stiegler writes that digital technology "creates a new organization of the circulation of the symbolic. Within this new mode of organization, suddenly the production of the symbolic becomes industrial, subject to industrial processes. Here you encounter the production of symbols on the one hand, and the consuming of such symbols on the other—an aporia because it is impossible to consume a symbol. The symbol is not an object of consumption; it is an object of exchange, of circulation, or of the creation of circuits of trans-individuation. So this situation suddenly produced what I call short-circuiting—of trans-individuation" (Stiegler and Rogoff 2009).

This systematic view sees retentions and protentions as circulations that are subject to control, manipulations that add political and economical considerations to digital objects; on the other hand, it also implies a reconsideration of the position of subjects and objects. Digital objects together with algorithms become the control of retentions (which can be short-circuiting and long-circuiting); the subject that contemplates natural objects or operates technical objects in factories or workshops could articulate causalities of perceptions and now becomes a processor of information. This approach takes the investigation of a classical question about cognition and AI and transforms it into social and political questions.

The second question concerns relationality, which is closely related to the first question, yet it is rather a metaphysical one than a political one.

And by digital objects, I want to propose here an opposition between the relationality and the substantiality. The key point at which a digital object differs from a technical object can be summarized as follows: A theory of digital objects demands a synthesis between Simondonian individualization and the Heideggerian interpretation of ready-to-handness (Heidegger would reject the idea that Simondon's thesis regarding technical objects poses any ontological questions, while Simondon would very much like to separate the technical from the social).[8] In the case of Simondon, in a mechanical system the contact point is the action of the relations—for example, the physical contacts between wheels and pulleys, the flow of electrons in electronic devices such as diodes. The relations that were once in a physical form are now turned into another material form, which is code or data. What was intangible before now can be made tangible and explicit, and be visualized in different forms. These relations are mobile and homogeneous. Data become objects and also the source of relations; this means the objects can join together materially through transmission networks, codes, and so on. The second point is that relations in Heidegger's technical objects are not material but temporal, since for Heidegger being can only be understood through time. The world is the spatiality that composes matrixes of relations, while these relations must be understood in a temporal sense, which Heidegger calls "care" (*die Sorge*). The problem with Heidegger is, how can we understand the new system of time with information machines that operate through digital objects? How can these two types of relation be understood in a technical system that is also digital?

We see first of all that these digital objects are also programmable, and they are themselves in the process of becoming computer programs. Relationality is the point where algorithms act, and at which data are related to each other. The evolution of technical standards from GML to XML to Web ontologies blurs the distinction between a simple text file and a structured computer programme. One can rewrite the whole code of a digital object, change its identity, and delete it in a second: what, then, is the substance of a digital object when its nature and identity are totally changed from point A to point B? One has to go down to the level of signals and voltages, but as we saw in the previous paragraphs, at that level objects become inconceivable. The question of substance proves bankrupt here. The problem of substance reveals the collapse of a universal monism. The transcendence of the object thus totally collapses in digital objects. In technical objects, we already encounter this problem, since these objects are

---

[8] Jacques Ellul quotes Simondon: "It is the ensemble, the interconnection of technologies, that makes this both natural and human polytechnical universe. . . . In existence, for the natural world and for the human world, the technologies are not separated, because there does not exist a thinking developed highly enough to permit theorizing about this technical network of concrete ensembles. . . . Beyond technical determinations and norms, we would have to discover polytechnical and technological determinations and norms. There exists a world of the plurality of technologies, with its own peculiar structures" (Ellul 1980, 82).

man-made objects, but we can still insist on the substance of the material, the perfectness of mathematical formulae, and so on. With digital objects, the transcendent aspect is further weakened, since virtually anyone can make and destroy these objects by pressing a key on the keyboard or clicking a mouse. In what is called the technological form of life, we are witnessing the flattening of the transcendent,[9] and objects fall into the field of total immanence. A new theory must therefore move away from the question of substance, and that for me is a theory of relations.

This chapter far from fully demonstrates a philosophy of digital objects; the investigation it proposes covers only a small part of the research I have done in recent years. It serves as an open invitation to engage with the philosophy of the Web and a phenomenology of digital objects. But for a philosophy of the Web to exist at all, one must move beyond the engineering principles and architecture of the Web itself, though one must always fully bear them in mind. For philosophy is not a representation of reality but reality itself, not one that gains its meaning from the mind of a thinker but one that comes out of the minds of thinkers through the significations of the world. We can certainly envision the expansion of the Web and future "breakthroughs" of technologies, but though a philosophy of the Web is on its way, it will never attain fullness without a theory of digital objects.

References

Aristotle. 1956. *Metaphysics*. Edited and translated by John Marrington. London: Everyman's Library.

———. 1984. *Categories*. Translated by J. L. Ackrill. In *The Complete Works of Aristotle*, ed. Jonathan Barnes. Princeton: Princeton University Press.

Berners-Lee, Tim. 2001. "Semantic Web." *Scientific American* (May). At http://www.scientificamerican.com/article.cfm?id=the-semantic-web

Clark, Andy, and David Chalmers. 1998. "The Extended Mind." *Analysis* 58:10–23.

Dretske, Fred. 2004. "Knowing What You Think Vs. Knowing That You Think It." In *The Externalist Challenge*, ed. Richard Schantz, 389–400. Berlin: De Gruyter.

Ellul, Jacques. 1980. *The Technological System*. New York: Continuum.

Floridi, Luciano. 2009. *Against Digital Ontology*. At http://www.philosophyofinformation.net/publications/pdf/ado.pdf

Gibson, William. 1984. *Neuromancer*. New York: Ace Science Fiction.

---

[9] Scott Lash (2002) uses "transcendental" instead of "transcendent." Here I follow the distinction made by Kant: transcendental means *a priori*, which doesn't need empirical evidence, while transcendent means an effect of supposed experience that exceeds the cognitive faculties of human beings.

Haugeland, John. 1993. "Mind Embodied and Embedded." In *Mind and Cognition: 1993 International Symposium*, ed. H. Yu-Houng and J. Ho Houng, 121–45. Taipei: Academica Sinica.

Heidegger, Martin. 1967. *Being and Time*. Translated by John Macquarrie and Edward Robinson. Oxford: Blackwell

———. 1988. *The Basic Problems of Phenomenology*. Translated by Albert Hofstadter. Indianapolis: Indiana University Press.

———. 2001. *Zollikon Seminars: Protocols, Conversations, Letters*. Translated by Franz Mayr and Richard Askay. Evanston: Northwestern University Press.

Kant, Immanuel. 1996. *Critique of Pure Reason*. Translated by Werner Pluhar and Patricia Kitcher. Indianapolis: Hackett.

Lash, Scott. 2002. *Critique of Information*. London: Sage.

Rotenstreich, Nathan. 1974. *From Substance to Subject: Studies In Hegel*. The Hague: Nijhoff.

Simondon, Gilbert. 1980. *On the Mode of Existence of Technical Objects*. London, Ontario: University of Western Ontario Press.

———. 2005. *L'invention dans les techniques: Cours et conferences*. Paris: Seuil.

Stiegler, Bernard. 1998. *Technics and Time*. Volume 1. Stanford: Stanford University Press.

———. 2009. *Technics and Time*. Volume 2. Stanford: Stanford University Press.

Stiegler, Bernard, and Irit Rogoff. 2009. *Transindividuation*. At http://www.e-flux.com/journal/view/121

CHAPTER 5

WEB ONTOLOGIES AS RENEWAL OF CLASSICAL PHILOSOPHICAL ONTOLOGY

PIERRE LIVET

1. Introduction

Philosophical ontologies, when they are seen as giving a definition of the fundamental types of entities that there are in our world—and in possible worlds—are nowadays mixtures of Fregean and either Aristotelian or Ockhamian conceptions. Everybody has in mind the Fregean distinction between reference (*Bedeutung*) and meaning or signification (*Sinn*), and their possible relation with argument and function in logic and mathematics (Frege 1984). Fundamental ontological entities have to be coordinated with that duality. In an updated Aristotelian tradition, substances taken as substrates (particular substrates, since universal substances are defended only by a few philosophers) are the targets of the operation of reference, and properties (universal and particular ones) are the ontological counterparts of predicates, themselves partially related to functions (Grenon 2003). In a similarly renewed Ockhamian tradition (the so-called tropist current), referents are identified by the compresence of particular qualities or concrete properties, and signification is related to similarities between particular qualities or properties. In a more classical nominalist current, reference is made to particular substrates, and signification implies the contribution of particular properties, universal properties being excluded (Livet and Nef 2009).

At first sight, present Web ontologies seem to be combinations of addresses (URI, URL) and names of classes in different domains, with the help of some basic relations like "has a link with," "is a . . . ," and maybe "is a part of . . ." that make it possible to build networks representing the ontology of a domain, and the articulations between different domains (Munn and Smith 2008). Addresses could be seen as the correspondents of referents, and names as the correspondents of significations.

In what follows, I suggest that upon closer examination we discover that Web ontologies introduce more subtle distinctions—at least a triadic distinction instead of a dualistic one. I also suggest that we have to conceive Web ontologies dynamically. In the dynamics of the construction and the evolution of Web ontological networks, the ontological type

Philosophical Engineering: Toward a Philosophy of the Web, First Edition. Edited by Harry Halpin and Alexandre Monnin. Chapters © 2014 The Authors except for Chapters 1, 2, 3, 12, and 13 (all © 2014 John Wiley & Sons, Ltd.). Book compilation © 2014 Blackwell Publishing Ltd and Metaphilosophy LLC. Published 2014 by Blackwell Publishing Ltd.

of an item will not really remain fixed but can change in relation to its interactions with the transformation of the structure of the network. In a sense, everything in Web ontology is an address, and the difference between types and addresses (between classes of content and references) is only given by different modes of cross-relation: chains of relations between addresses, and transversal relations across these chains for types. The mode of structuration is more fundamental than the kinds of entities. This relational and dynamic approach has to be taken more seriously by philosophical ontologies, which are usually centred on static entities.

In this dynamic perspective, the old dualities of philosophical ontology could be reduced to the duality of two operations: distinguishing items—trying to identify them, even only relatively—and making them and their articulations explicit. For the sake of brevity, I will use the neologism "explicitation" instead of "making explicit." Explicitation is a process that starts with a basic operation of distinction that distinguishes items that were previously undistinguished and proceeds by also using other previously undistinguished items in order to make explicit the articulations between the distinguished items. The definition of the type of the entities is then relative to the stage of the process of explicitation. Because distinctions are coarse-grained ones at the beginning of the processes, several kinds of entities are still not distinguished at this stage (for example, the considered items could indifferently be a particular substrate, a particular quality, or even a universal property). A specific ontological type of entity is defined in relation to the stage at which it has been necessary to distinguish it from other previous ontological types of entities. Up to this stage, the ontological type of an entity can be considered a "floating type," floating between the different types that are still possible.

In the Web the ontology and its explicitation are conditional on the process of developing the network and are related to the different stages of this process, considered a recursive one. In this sense, the Semantic Web (Web 3.0)—the kind of Web for which ontologies are required, even if they are also useful for the knowledge Web (Web 4.0)—could retain some features of the social Web (Web 2.0) in which folksonomies emerge from the free tagging of information and objects by users when one tries to relate these tags to each other. This dynamics of recursively making explicit floating types seems to be a fundamental property of the ontologies of Web 2.0, 3.0, and 4.0, and could also be a property of a philosophical ontology that includes in it the ontology of its own processes of explicitation.

2. Addresses, Reference, and Signification

Addresses on the Web seem just to be pointers to locations. As pointers, they surely have the function of referring to items. Tags or names seem to have the function of pointing out meaningful content. Accordingly, they seem to be candidates for the function of signifiers in the domain of thought and for predicates in a logical language. Addresses and tags seem to correspond to the two main Fregean categories.

Unfortunately (or, as I will suggest, fortunately), this is far too simple. First, in the social Web, tags belong to folksonomies. If you want to move from those folksonomies to the Semantic Web, you have to solve the difficult problem of how to homogenize the different contents to which the tags refer. If you are successful in doing so, you have given up the functions of tags that were not reducible to conceptual contents: for example, contextual connotations and marks of reference to the group of the users of these tags. A way of saving a part of those very diverse functions is to take tags as labels for nodes in a network: at least some of the contextual functions will be kept. In this case, tags are no longer reducible to Fregean significations, for those are not context sensitive. Since pointing out nodes in a network is also a referential function, tags keep some referential function, but now this function is contextualized.

Similar complications occur for addresses. We could believe that addresses point to nodes, which are locations in the structure of the network. The problem is that any change in the network changes the structure of its relations and so also the relational role or the nodes. Some new nodes become accessible if you add links and of course new added nodes are reachable by these new links. As there is no previously existing homogenous structure of locations, no space pre-existing the nodes and the links of the network, the structure of the network space is changed each time new nodes and new links are added—or old nodes and old links are erased.

In a sense, this network structure makes both addresses and tags (as related to nodes) more similar to inferential paths. The inferential content of a conceptual term (for Frege a conceptual term is the counterpart of a signification in the realm of thought) is determined by all the paths of inference in which it participates. If you change the tree or the network of those paths, you change the meaning of the concept. Frege would have replied that even if conceptual representations and processes of psychological inference could evolve, the significations are stable. But adding new links is not only to add newly discovered but stable significations, it can in addition change the contextual import of the previous ones. The ontology of the Web has to take into account this possibility.

All this would be true even if addresses were only ways of pointing out locations, in a network that is not a homogenous and pre-existing structure but a structure that is constantly rebuilt. But addresses are not reducible to that function alone. They give access to resources. Giving such access is not only to point towards the resource. Giving access is already to use the resource as such. If giving access is considered a function, using the resource is in a sense a function (a function of the function of giving access). We are far from the strict notion of referent. Or, to take things another way, we discover that having access to a referent is much more than pointing it out.

Things are even more complicated. Monnin (2012) has drawn my attention to the fact that "resources" do not always give access to the same content (if the content is updated every day, for example). In this respect they behave more like indexicals, but in contrast to indexicals, they give you some access to the content (if there is one) without requiring additional information. In addition, an address is not just a URL (uniform resource locator) but more generally a URI (uniform resource identifier). URIs just give the name of a resource, independent of its location and the way in which it is referenced. In a sense, URIs could have been taken as the real Fregean referents, because Frege takes them as referents independently of any particular way of referencing. But from another perspective, URIs are the names of the referents, as they are not what is there in the indicated location. Are they proper names or common ones? They could be seen as the common names (common to different ways of referencing) of proper names. Here again the ontological resources of the Web ontologies pass the Fregean distinctions by recombining features of their different categories.

Another interesting distinction of Web ontologies is the difference between the closed world assumption and the open world assumption. The open world assumption implies the possibility that our ontology is incomplete. New ontological types could need to be added in the future, contrary to the requirement of the closed world assumption (according to which all types of entities have already been spelled out). Since reference is now related to access, this distinction enriches the list of the different functions of referring. We can refer to an accessible referent by its proper name, knowing its location. We can refer to an accessible referent without knowing its location, but being able to name it and to put it on our list of referred entities. Eventually we can refer "out of the blue" to referents for which we have neither location nor name. They are only supposed to be accessible if we can extend our world, or go beyond the present world to which we currently have access. The notion of "referent" can be taken in each of these three ways.

Apparently, this is not possible for significations. We have no real common names for entities of the third category, entities to which we currently have no access at our disposal. We indicate them by using paraphrases and what could be better called "indefinite" descriptions ("one of the entities to which we do not currently have access at our disposal"). Only a dualist distinction seems possible: significations can be accessible by common names, without a real definition of the precise content of the common name (only a nominal one), or by their conceptual content.

Is there a way to restore the symmetry between referents and significations, and to have three real distinctions for both? We could introduce the triadic distinction between (a) what is not distinguished, (b) what is distinguished but not made explicit, and (c) what is distinguished and made explicit. This distinction works both for referents and for significations. If "referent" is related to "distinguished," a referent "out of the blue" is "undistinguished," a distinguished referent is a referent to which we can point even without knowing its location; knowing this location implies relating it to other distinguished entities, and this consists in making its distinction explicit. In the same way, the signification of an entity that is not distinguished from other ones is "out of the blue"; the signification of an entity is distinguished but not made explicit when it can be captured and when the ways in which it distinguishes it from other types of entities (what we could call its distinguishers) are still not themselves distinguished; the signification of an entity is explicit when the ways of distinguishing it from other types of entities have themselves been distinguished.

Apparently this threefold distinction is an epistemic distinction, not an ontological one. But epistemic distinctions have ontological bases: the processes by which they proceed, and the structure of the reality that allows those processes to proceed. We could have entities the structure of which cannot allow us to distinguish them from other ones; entities the structure of which allows us to distinguish them from other entities but do not allow us to distinguish their distinguishers; and entities the distinguishers of which can themselves be distinguished.

3. Floating Types and Recursive Process of Explicitation

The triadic basis gives rise to a recursive process, when it is possible to distinguish entities by using other entities to distinguish them, then to make explicit the previous distinctions by distinguishing the distinguishers (using other distinguishers), then to make explicit the distinguisher by distinguishing these other distinguishers, and so on.

For example (if we use the classical types of ontological entities) substrates can distinguish themselves from other substrates: particular substrates use other particular substrates as their distinguishers. Particular qualities can use other particular qualities as distinguishers. But how can particular substrates be distinguished from particular qualities? If we say that substrates as such distinguish themselves from qualities as such, using qualities as distinguishers, this answer gives rise to another question: What are the features of substrates and qualities that distinguish them from one another? We could answer by using different kinds of relations or even new types of entities. For example, if "quality"—and not "substrate"—is the most basic category, the relation of compresence of qualities (a quality is compresent with another if they co-exist together in a bundle) builds a substrate with its qualities (at the same level as the qualities and the relation). If we admit a difference between two kinds of entities (substrates and qualities), we have introduced not only the relation of difference but also, at another level, kinds as types of new entities—or maybe universals.

We see that when in the first step we focus on substrates, we assume at least other substrates in order to distinguish one substrate, but those substrates, used as distinguishers, are not themselves distinguished. They are still undistinguished. In this step we have considered three possibilities: the first is that there are only substrates, the second is that we could have only qualities and one relation—if the so-called substrates require the compresence of qualities—and the third is that we have only relations—if we consider compresence and similarity (or difference) as more fundamental than their relata. In the second step (differentiating substrates from qualities, for example) we could have kinds or universals, or relations, or both.

Suppose that in a third step we have to make explicit as distinguishers what is the glue or connection articulating one substrate and its quality. We have then to make explicit one distinguisher of this substrate and its quality, the relation between the substance and its quality. If the quality is a universal, the substrate exemplifies this universal, and the relation is one of exemplification. If the quality is a particular, we need in addition to the relation of compresence the relation of instantiation, as a particular compresence is an instantiation of the universal relation of compresence.

If we characterize the ontological entities that we have introduced by the step in which they have been made explicit, we see that particulars as referents are entities of the first step (we could use particular substrates or particular qualities instead, but then qualities as particulars would be used only as identifiers). Qualities differing by their type from substrates are then entities of the second step. Relations distinguished from qualities and substrates—here we need to use as distinguishers universals, a category that is still not made explicit—are entities of the third step. Probably kinds or universals, if we use them, are entities of fourth step, and so on.

Such a recursive development makes it possible to take into account from the beginning dynamic processes in ontology, instead of having to presuppose static entities defined once and for all. The development makes new distinguishers accessible. Along the way, it allows new distinctions. We progress from a very coarse-grained step toward other, finer-grained steps. In a sense, in a very first step, a zero step, we have only undistinguished entities. Then in step 1 we have particulars. We do not know whether they are substrates, qualities, relations, or exemplifications or instantiations of universals. In step 2 we distinguished among particular qualities and substrates. Relations, if they are in use, are still not distinguished and will be in a third step.

Let us go back to Web ontologies. Our dynamic process is a presentation of ontology that is more natural than the classical static ontologies, and is needed in order to characterize the development of ontologies in the different types of Web. Addresses can be used at first in a coarse-grained way. In this first step there are distinguishers, but they themselves are still not distinguished: their different possible operations—finding the location, naming the referent, relating it to other nodes of the network, making possible associations and inferences, giving access to stable contents or to continuously updated contents—are still taken together and are not necessarily distinguished from one another. Tags are also distinguishers, needing the machinery of URIs and URLs in order to be related to one location and then to each other. They are distinguished from addresses since they can use common names as well as idiosyncratic ones. We should note also that different tags can refer to the same location. When different tags are attached to the same location, they work as particular qualities attached to the same substrate or related by compresence. But if one tag can be used in different locations without equivocity, it has the status of either a relation or a universal. Tags seem to be entities emerging in our second step. Of course, in order to make explicit how addresses, tags common to different addresses, and tags common to the same addresses are organized, we need to specify the relations that build the network and avoid equivocity. These relations are made explicit in a third step.

For sure, in order to build the network of addresses, Web engineers have had to define relations from the beginning. But if for these engineers those relations have been made explicit, they do not function on the Web as made explicit—on the contrary, for users they are the hidden and undistinguished part of the Web. Relations become distinguishable for the user only with the social Web and the Semantic Web. The ontological entities that the users of the Web can refer to, using the technical facilities of the Web, seem to follow a progression very similar to the development of the process of explicitation, the outline of which I have just presented. The task of the engineers of the different kinds of Web is to build structures that allow the development of dynamics that can fit with this rather "natural" development of the explicitation of the kinds of ontological entities. This development is a progressive one; formally it can be recursive. We can use a kind of entity as a distinguisher without making explicit what this kind is, and we do not make explicit a kind of entity except if we need to specify its conditions of use. Such an explicitation needs other but not yet made explicit kinds of entities, and so on.

4. Conclusion: Points Still to Be Made Explicit

I hope to have shown here that a conception of ontology is possible that presents a fundamentally dynamic aspect, more in tune with the dynamics of the extension and the emergence of different forms and functions of Web 2.0, 3.0, and 4.0. The condition for doing so is to take into account in the ontology the processes by which we distinguish entities that were undistinguished at first and then make explicit those distinctions by distinguishing the entities that we have only used as distinguishers in our first task.

In this process, the distinction between reference and signification could seem to have been blurred, because referents and significations are both distinguishers. But it is not really the case, for in our first step (when a substrate uses another substrate to be distinguished) we have used a distinction between entities that are supposed not to differ by their kind, so that this distinction between several substrates was only used to distinguish a particular referent. Significations are at stake in the further steps—with qualities, relations, and universals.

How is this distinction maintained in the operations of Web ontology? We have seen how addresses and tags—our candidates at the basic and coarse-grained levels (coarse-grained with respect to ontological sophistication) have cross-functions. Addresses allow chains of paths from one location to another, and tags can either subsume different nodes or specify different features of the same nodes—while these features can be also put together under the umbrella of a generic tag. The knowledge Web requires a more elaborated and sophisticated ontology, with relations and processes of inference, for example, but it could be regarded as taking place in the same overall dynamic process of recursive ontological explicitation.

The parallel between this ontological development and the new functions built into the Web could lead us to define the canonical ways of such a development and to give in this way ontological landmarks that would be useful for comparing things and avoiding becoming lost in the complexity of the Web. But such a reflexive sophistication will also introduce more constrained specifications that could be obstacles for the creativity and the imaginative extensions—related to the open world assumption—that are attractive in the different kinds of Web. It seems possible to deal with those difficulties by using the process of ontological development not only in an upward manner, from the coarser-grained levels to the more sophisticated ones, but also in a downward manner, coming back from explicitation to indistinct ontological stages—a way of being again more liberal with the definition of ontological categories—as all these levels have their own ontological relevance.

But here the purposes of philosophy and the Web diverge, since the philosophical return to the first levels of explicitation is at the same time a backward move from sophisticated propositions to the most basic and fundamental entities, while the same move within the Web framework would be a way to relax unduly restrictive formal constraints and to open fresh potentialities still not explored by users.

References

Frege, Gottlob. 1984. *Collected Papers on Mathematics*. Edited by Brian McGuinness. Oxford: Blackwell.

Grenon, Pierre. 2003. "BFO (Basic Formal Ontology) in a Nutshell: A Bi-Categorial Axiomatization of BFO and Comparison with Dolce," IFOMIS Report 06_2003.

Livet, Pierre, and Frédéric Nef. 2009. *Les êtres sociaux*. Paris: Hermann.

Monnin, Alexandre. 2012. "The Artifactualization of Reference and 'Substances' on the WEB. Why (HTTP) URIs Do Not (always) Refer nor Resources Hold by Themselves." *American Philosophical Association Newsletter on Philosophy and Computers* 11, no. 2.

Munn, Katherine, and Barry Smith (eds.). 2008. *Applied Ontology: An Introduction*. Frankfurt: Ontos.

Smith, Barry. 1984. "Ontology (Science)." In *Formal Ontology in Information Systems*, edited by Carola Eschenbach and Michael. Gruninger, 21–35. Proceedings of FOIS 2008. Amsterdam: ISO Press.

CHAPTER 6

BEING, SPACE, AND TIME ON THE WEB

MICHALIS VAFOPOULOS

Introduction

The fast-growing and highly volatile online ecosystem of the Web is an integral part of everyday life, transforming human behavior in unexpected ways (Vafopoulos 2006). Traditional borderlines between public and private, local and global, and physical and digital have started to blur. New types of property, identity, and market have emerged. Which parts of the existing scientific armory can help us to enrich our understanding as far as the Web is concerned, and to build models that anticipate positive future developments? To address such issues we need to overcome the common metaphors used in the scientific modeling of collective phenomena that can be summarized either as a society encompassing individuals or as an "invisible hand" generating social optimums out of self-contained rational agents or auto-organized collective actions (Latour 2011).

In this chapter, we engage with two brands of philosophical thinking that most influence the proposed metaphors. First is the Actor-Network Theory (ANT) hypothesis "not to reduce individuals to self-contained atomic entities but let them deploy the full range of their associates" (Latour 2011, 806) that the Web makes possible due to the plethora of user datascapes. Second is the archetypal view of time by Bergson, who originated the primordial role of durations. Web users' visiting sessions can be better analyzed as durations, evaluated heterogeneously according to atomic memory and consciousness. In their day Bergson and Tarde (whose work inspired ANT) lost scientific debates with regard to the prevalent notions of time and society, respectively, but for both of them the Web offers grounds for belated revenge on the arguments of their antagonists. At the 1922 conference of the Société Française de Philosophie Bergson failed to persuade the scientific community with his arguments that time is a succession of indeterministic and heterogeneous durations that are irreversible and subject to memory and consciousness (Bergson 1922 and 1965). Einstein's mathematical proof concerning deterministic physical time prevailed due to the lack of a sound scientific riposte. The significance of visiting durations on the Web highlights Bergsonian time as the "time of social systems."

Philosophical Engineering: Toward a Philosophy of the Web, First Edition. Edited by Harry Halpin and Alexandre Monnin. Chapters © 2014 The Authors except for Chapters 1, 2, 3, 12, and 13 (all © 2014 John Wiley & Sons, Ltd.). Book compilation © 2014 Blackwell Publishing Ltd and Metaphilosophy LLC. Published 2014 by Blackwell Publishing Ltd.

Durkheim, who was Bergson's classmate at the École Normale Supérieure, became one of the dominant figures in social research by advocating that society is superior to the sum of its parts, and he initiated two levels of study between individual psychology and a sui generis societal approach (Latour 2002). In contrast, Tarde argued that individuals and society are two sides of the same coin and should not be analyzed separately. The atomic profiles of personal data that are massively collected through the Web offer strong empirical evidence for modeling actor and networks as a single entity. Tarde's arguments have started to have preponderance over Durkheim's, mainly through the propagation of ANT across scientific borders. On the Web Bergsonian time meets the Tardian approach to individuals and society. The reconstruction of atomic memory and consciousness through the analysis of visiting sessions on the Web emphasizes the reciprocity of actor and networks in understanding human behavior.

In the first section below I indicate the research questions under investigation in this chapter, and the main results. In the next section I analyze being, time, and space. I also provide original definitions of beings, Web beings, Web time, and Web space. In the third section I discuss how the emergence of the Web changes the notion of existence, space, and time in general. In the final section I consider some further implications of the proposed conceptualizations.

1. Research Questions and Main Findings

Monnin's call (Monnin 2011) for a more precise description of Web "resources" and Halpin's ontological definitions (Halpin and Presutti 2009a) acted as a good stimulus for investigating the basic constituents of existence on the Web. The present chapter addresses two research questions:

  I. What are the main characteristics of being, space, and time on the Web if it is considered a self-contained system that exists in and by itself?
 II. How do these idiosyncratic features of the Web transform the traditional conceptions about actual being, space and time?

Web beings are defined as beings that can be communicated through the Web. The URI (uniform resource identifier) is the minimal description of invariant elements in this communication and acts as the "fingerprint," the "interlocutor," and the "borderline" of Web beings. It enables direct access as well as the exportation and importation of content. This linking capability, combined with the digital nature of Web being, constitutes the notion of virtualization. Virtualization jointly describes the augmented potentialities of Web being as a digital and distributable unity. The digital facet is distinguished from the physical by five characteristics, namely: nonrivalry, infinite expansibility, discreteness, aspatiality, and recombination. The Web, as a distribution network, restricts nonrivalry and infinite expansibility but extends aspatiality, atemporality, and recombination (Vafopoulos 2012).

Location in Web space is specified by the Web being's URI and the URIs of incoming and outgoing links. These links provide orientation by acting as a three-dimensional "geographic coordinate system" on the Web. Web space is heterogeneous because Web beings are not characterized by the same network microstructure.

What information consumes is attention. Attention on the Web could be approximated by the visiting time in a Web being, that amount of time a user spends at a particular resource. The primordial role of visiting durations in the Web's existence qualifies Bergsonian time as the best candidate for modeling Web time. The Bergsonian approach to time as successive durations implies indeterminism, heterogeneity, and irreversibility. The new property that the Web contributes to the context of "social time" is that it enables the automatic and effortless recording of a starting time and an ending time for all visiting sessions. Durations are becoming discoverable, observable, traceable, processable, and massive.

The Web transforms spatiotemporal actuality by adding flexibility and an enriched set of choices for human action. Web space tends to both expand and limit the notion of physical space. Physical space becomes more discoverable and traceable. Transportation and transaction costs become negligible, creating a new range of potentialities. Some of the human activities through or on the Web are available asynchronously, in part synchronously, and continuously. Networked individuals can mobilize part or the whole of their communication system, operating in a flexible, less-bounded, and spatially dispersed environment. The resulting issues are related to the self-determination of Web users and the prerequisites for the Web to be an open platform for innovation and participation.

2. Existence in Web Space and Time

2.1. The Facets of the Web

Although the Internet was introduced twenty years earlier, the Web has been its "killer" application, with more than two billion users worldwide accessing some one trillion Web pages. Searching, social networking, video broadcasting, photo sharing, and microblogging have become part of everyday life, and the majority of software and business applications have migrated to the Web. The Web is evolving from a simple fileserver to an enormous database of heterogeneous data. The fundamental hyperlinking property that enables positive network effects in document sharing (Web 1.0) is rapidly expanding to social spheres, contexts (Web 2.0), and

URI-based semantic linkages among data (Web 3.0) (Vafopoulos 2011a and 2011b).

In this chapter I use the term "Web" to describe four interconnected parts, namely: Internet infrastructure, Web technologies, online content, and users. The Web was initiated as software for codes of interlinked hypertext documents accessed via the Internet. Using a browser, users access Web pages that may contain text, images, videos, or other multimedia and navigate among them using hyperlinks. The Web constitutes an information space in which the items of interest, referred to as resources, are marked up by a set of rules (HTML, or hypertext markup language), identified by global identifiers (URIs), and transferred by protocols (HTTP, or hypertext transfer protocols). They could be considered "contemporary forms of what the Greeks of antiquity called hypomnemata" (i.e., mnemotechnics) (Stiegler 2010). The Web has become the most successful and popular piece of software in history because it is built on a technical architecture that is simple, free or inexpensive, networked, based on open standards, extensible, tolerant to errors, universal (regardless of the hardware and software platforms, application software, network access, public, group, or personal scope, language, cultural operating system and ability), powerful, and enjoyable.

Human participation on a massive scale upgraded the Web from a piece of software to a "social machine" (Hendler et al. 2008). The Web has evolved into a highly interdependent living ecosystem, which affects users and nonusers in everyday, and sometime life-critical, choices. During the past decade the transformative power of the Web stimulated its study as a stand-alone techno-social ecosystem under the umbrella of Web science (Berners-Lee et al. 2006; Vafopoulos 2011c).

2.2. Being on the Web

The notion of "resource" has been selected to describe the underpinning unit of the Web ecosystem. Sometimes it is interchanged with terms like "object" and "thing" (and others). Basically, the concept of resource has strong economic and ecological connotations. In economics, there are three basic resources: land, labor, and capital. The term "resource" enjoys twelve appearances in the scientific classification of economics.[1] In a broader sense, they are human, natural, and renewable resources. In addition, in the relevant Wikipedia entry on resource there is not a single word about the Internet or the Web.[2] On the other hand, terms like "object," "entity," and "thing" are too general and not descriptive enough, since

---

[1] See http://www.aeaweb.org/jel/jel_class_system.php
[2] See http://en.wikipedia.org/wiki/Resource

they convey multiple meanings in different contexts. Thus, a new concept for the Web's cells is required to demonstrate the coevolution of new technological and social aspects; the living nature of the Web consists in internal laws of existence, dynamism, and transformational impetus. It is important to introduce existence on the Web with a clear, minimal, and pragmatic definition that describes the Web as an integral part of the world and that could be useful for Web scholars and engineers from diverse origins.

This effort is initiated by baseline definitions of being in general and Web being in particular.

> *Being:* A being exists if and only if there is a communication channel linking to it.

This definition is not only practical but also general enough to include many existing theories in philosophy. For instance, it could be considered an alternative and minimal approach to "Dasein" as reintroduced by Heidegger in 1926 (Heidegger 1962). As Latour explains, "Dasein has no clothes, no habitat, no biology, no hormones, no atmosphere around it, no medication, no viable transportation system. . . . Dasein is thrown into the world but is so naked that it doesn't stand much chance of survival" (Latour 2009, 140).

The concept of "communication channel" in defining existence encapsulates the core idea of ANT about "circulating entities" and the useless dichotomy between individuals and society that is commonly the case across various disciplines (Latour 2005). In the Web era, "to have" (e.g., friends, connections, identities) becomes of equal importance to "to be" and should be studied along with the various manifestations of "to be." One aspect of being on the Web is that the communication channel is more concrete, identifiable, and visible. The URI is the fundamental technology in creating communication channels on the Web.

URIs[3] (e.g., http://www.vafopoulos.org) are short strings that identify resources on the Web: documents, images, downloadable files, services, electronic mailboxes, and other resources. They make resources available under a variety of naming schemes and access methods, such as HTTP and Internet mail addressable in the same simple way. They reduce the tedium of "log in to this server, then issue this magic command . . ." to a single click. Every URI is owned by a physical or a legal entity, which has the right to sell it or to provide access to it by any other entity (see Halpin and Presutti 2009b for a thorough discussion of identity and reference on the Web).

---

[3] See http://www.w3.org/Addressing/

Web beings: Web beings are beings that can be communicated through the Web.

The URI is the minimal description of invariant elements in communication through the Web and unambiguously characterizes the Web being. It is like the "fingerprint" of the Web being because is directly connected to existence (birth, access, navigation, editing, and death). All the other characteristics of Web beings may change over time (e.g., appearance, content, structure), but a change in URI means the death of an existing Web being and the birth of a new Web being. The main role of URI is to categorically identify the Web being and to enable navigation and recombinations across the network of Web beings.

Web beings can be information-based or non-information-based (as resources defined in Halpin and Presutti 2009a), but in both cases they are named, referred, and identified through at least one exclusive URI, which constitutes the minimum amount of digital information necessary for the aforementioned operations.

The question is how a being can be artifactualized (Monnin 2009) into a Web being. In particular, how is digitality mutated by the linking potential, enabling beings to be anywhere at any time?

2.3. Virtualization = Digitality + Linking

The Web revolution is based on the realization of publishing, interconnecting, and updating digital objects globally at low cost and with little effort. The massive reallocation of action over distributed datascapes takes place because the digital is augmented and ramified by linking capacities. Intuitively, every user can access all available Web beings anytime from anywhere. We are all "potential" (or "quasi") owners of each Web being, in the sense that it may not reside in our memory device but can be downloaded almost instantly. This fundamental expansion of existence can be better captured by the concept of "virtualization." Virtualization jointly describes the augmented potentialities of a Web being as a digital and a distributed unity. Linking is based on digitality but is metamorphosing it into a more volatile, open, and multifunctional artifact. According to Lévy, "Virtualization is not derealization (the transformation of a reality into a collection of possibles) but a change of identity, a displacement of the center of ontological gravity of the object considered. . . . The real resembles the possible. The actual, however, in no way resembles the virtual. It responds to it. . . . Rigorously defined, the virtual has few affinities with the false, the illusionary, the imaginary. The virtual is not at all the opposite of the real. It is, on the contrary, a powerful and productive mode of Being, a mode that gives free rein to creative processes" (Lévy 1998, 124).

2.3.1. The Digital Aspect. The digital aspect of existence has been extensively investigated by diverse research disciplines (see, e.g., Negroponte 1995 and 2000; Kim 2001; Castells 2003). Digital beings are beings that are constructed by sequences of bits, 0s and 1s. They are usually created by mental efforts and occupy negligible parts of physical space. By design, they cannot directly satisfy basic needs like food, water, shelter, and clothing. The most relevant analysis emerges if we focus on "digital goods," which are digital beings with economic value. Quah (2003) distinguishes digital from physical goods by five characteristics, namely: nonrivalry, infinite expansibility, discreteness, aspatiality (or weightlessness or spacelessness), and recombination. In particular, in economic terms digital goods are:

(1) *Nonrival.* Most physical goods are rival, in the sense that consumption by one consumer diminishes the usefulness of the goods to any other consumer (e.g., if I drink a bottle of water, I exclude everybody from drinking this particular bottle of water). In contrast to most physical goods, digital goods are nonrival, in the sense that many users can consume, say, videos and philosophical theories without preventing their use by others. The concept of nonrivalry plays an important role in economic analysis (Rivera-Batiz and Romer 1990; Shapiro and Varian 1999).

(2) *Infinitely expansible.* According to Pollock (2005), "A good is infinitely expansible if possession of 1 unit of the good is equivalent to possession of arbitrarily many units of the good—i.e. one unit may be expanded infinitely. Note that this implies that the good may be 'expanded' both infinitely in extent and infinitely quickly." Only in the limit case is perfect nonrivalry equivalent to infinite expansibility. Infinite expansibility and virtually zero-cost copying and distribution of digital beings are strong forces of change in the business model of media industries.

(3) *(Initially) discrete or indivisible.* Digital goods are (initially) discrete, in the sense that their quantity is measured exclusively by integers, and they are only instantiated to a quantity of 1. Alternatively, digital goods are not divisible. As Quah (2003, 17) explains, "Making a fractional copy rather than a whole one, where the fraction is distant from 1, will destroy that particular instance of the digital good." Indivisibility can be further refined by the properties of fragility and robustness. In economic terms, a digital good is robust if a sufficiently small and random reassignment or removal of its bit strings does not affect the economic value of the good (e.g., MP3 compression). Thus the digital good is defined as fragile (e.g., computer software).

(4) *Aspatial.* Digital goods are aspatial in the sense that they are everywhere and nowhere at once. The aspatiality of digital goods is not identical to the definition of aspatiality in Plato's theory of Forms or theory of Ideas (Ross 1953), which refers to the absence of spatial

dimensions and thus connotes no orientation in space. Digital goods are real bits located in physical devices (e.g., hard disks).

(5) *Recombinant.* Unlike the majority of ordinary goods, digital goods are recombinant. "By this I mean they are cumulative and emergent—new digital goods that arise from merging antecedents have features absent from the original, parent digital goods" (Quah 2003, 19).

2.3.2. The Linking Aspect. The characteristics of digital goods described above can proliferate through a compatible distribution channel, which is compatible with their prototypes, exchange, and access. The most efficient channel has proven to be the Web network. Within just a few years the Web has evolved into an enormous repository and distribution channel of information. Web technologies provide the technical platform for representing, identifying, interconnecting, and bartering addressable information. On the Web, the digital becomes tangible, editable, uniquely definable, and compatible with almost any format.

A cornerstone in this process is URI technology. The URI is the specific part of digital information contained in a Web being that identifies it as unique and enables direct linking and transfer to other Web beings. Every single URI identifies one Web being exclusively. Although each Web being has one generic URI, it can be identified through many other URIs. The URI is the "fingerprint," the "borderline," and the "interlocutor" of Web beings. It facilitates the "teleportation" of navigators (i.e., direct access) as well as the automatic exportation (e.g., through RSS, or really simple syndication, feeds) and importation of content from other Web beings (e.g., the inclusion of twitter streams in personal pages). The latter emerges, in the most concrete way, as a trade-off between original creation and reference to existing work. An old dilemma, which exists mainly in scientific research, returns as a practical question in creating and commercializing Web beings.

The Web, as a distribution channel, restricts nonrivalry and infinite expansibility of digital goods because of its limited concurrency capacity and costs imposed by the underlying infrastructure. Digital goods in the Web are initially discrete and indivisible like their underlying digital goods. Their distinctive characteristic is the facilitation of massive recombination and consumption in microchunks. In the current Web 2.0 era, users can easily edit, interconnect, aggregate, and comment on text, images, and video on the Web. Most of these opportunities are engineered on a distributed and self-powered way. Conjointly with massive information aggregation and recombination, the Web extends aspatiality and atemporality to digital goods. Distributed digital information is characterized not only by the fact that it enjoys low transportation cost but also by its accessibility anytime from anywhere. Actually, the Web expands aspatiality and atemporality from the local level (e.g., a personal hard disk) to the global level (e.g., a cloud disk).

2.4. Web Space Is the Online Network

Every Web being can be accessed directly by typing a URI in browser's address bar or indirectly through search engines or clicking on referral Web beings. Analogously to the physical world, Web space can be considered a division of the position and place of online content, created by the links among Web beings. Each Web being occupies a specific locus on the Web. The identification in Web space is given by the URI namespace, whereas the location is specified by a triplet of URIs: the Web being's URI and two sets of URIs: incoming and outgoing links of the Web being. These links provide orientation by acting as a three-dimensional "geographic coordinate system" for the Web.

Web space is heterogeneous because Web beings are not characterized by the same network microstructure. For instance, vafopoulos.org and w3c.org have different sets of incoming and outgoing links, and thus different pathways and possibilities to be accessed. The study of network structure at the micro- and macrolevel is an important issue in many disciplines, such as mathematics and physics, computer science, network economics, game theory, and social studies (see, e.g., Newman 2008, Jackson 2008, Albert and Barabási 2002, Watts 2003). Network metrics (e.g., hubs and authorities, betweenness, and eigenvector centrality) and algorithms (e.g., community detection) under different criteria calculate the "gravity" and the relative "distance" between beings in Web space. Likewise, eigenvector centrality was the initial base of Google's search algorithm (Page et al. 1998) that successfully assists us in exploring Web space and in locating Web beings.

How does Web space evolve over time? The creation or the deletion of a single Web being or link constitutes a change in Web space. Consequently, the evolution of Web space is fully described by the birth and death processes of Web beings and URIs.

2.5. Web Time Is Bergsonian

2.5.1. Attention: The "Currency" of the Web. Navigation in Web space results in Web traffic. Web traffic is defined as the successive activation of URIs recorded in a Web being's log file. In order to prevent confusion, the convention has been established that events on the Web are synchronized under the UTC (coordinated universal time). This is actually the first time that humanity has established a universal event log system to such an extent as a scalable and heterogeneous sociotechnical system. The Web's embedded logging system minimally records the access date and time, the user's IP (Internet Protocol) address, the viewed URI, the amount of transferred bytes, and verification of the status of the transfer.[4] For

---

[4] See http://www.w3.org/Daemon/User/Config/Logging.html#common-logfile-format

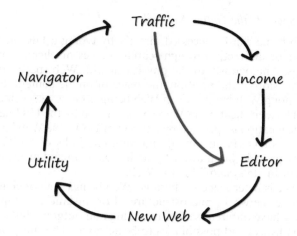

FIGURE 6.1. The inner function of Web economy. Navigators explore the Web to acquire utility, resulting in exploitable traffic for editors, who are motivated to update the existing Web.

analytical purposes, the initial logs are aggregated with respect to users or Web beings. In the first case, user sessions are formatted as temporally compact sequences of visits to a series of Web beings by a specific user. Alternatively, user sessions can be aggregated with respect to Web beings (for related issues in modeling Web traffic see Abdulla 1998).

According to the original architecture of the Web, the owner of a Web being has exclusive access to the log file, which describes the traffic of users in the local Web space. Lately, many companies have extended their control in Web space by collecting traffic data outside their Web beings (e.g., cookies). In most cases, this expansion takes place without the consent of users, raising important issues and debates about privacy and self-determination on the Web (Goldfarb and Tucker 2010; Angwin 2010).

In order to understand the central role of visiting durations in the Web's functionality, let us investigate the inner information flows, assuming that the Web is a self-contained system that exists in and by itself. Web users are navigating or/and producing Web beings. Thus, users are partitioned into navigators and editors of Web beings. Navigators explore the Web to acquire utility by consuming Web beings (see figure 6.1).[5] This navigation creates traffic streams for the creators of Web beings (editors). The main motivation for creating and updating Web beings is to attract visitors. Nonprofessional editors provide information, effort, and time for

---

[5] In Web economy analysis, Web beings can be specifically defined to be "Web goods." "Web goods" are Web beings that affect the utility of or the payoff to some individual in the economy (Vafopoulos 2012).

free, mainly for the sake of individual acclaim and reputation building, moral reward, and self-confidence. In some cases, the temporal segregation between effort and reward is explained by an expected increase in editors' opportunities for well-paid employment. In contrast, professional editors of Web beings are profit maximizers and take into account direct financial compensation (e.g., a blog with paid advertisements). In both cases, the time that visitors spend in Web beings (sessions) is crucial for their existence because it is directly related to editors' expectations. The resulting income acts as an incentive for professional editors to update the existing being and create new Web beings, producing a new Web space with novel possibilities for navigators to maximize their utility.

What information consumes is attention (Simon 1971). Attention on the Web is captured by the time spent visiting a Web being. The monetization of visiting sessions has created the largest internal Web market, with more than $63 billion turnover in 2010 (Yarow 2011). Hence, attention could be considered the "currency" of the Web and visiting durations the unit of exchange in the Web market.

2.5.2. Visiting Session: The Time of the Web. Apart from the "bookkeeping" clock time that is strictly defined by physics, time could be described as a series of choices in space. Analogously, Web time is a series of choices (visits) in Web space that can be better modeled as Bergsonian durations, because visiting selections attach semantic meaning and define casual relationships among Web beings.

The Aristotelian view asserts that time is like a universal order within which all changes are related to each other. Time is meaningless if there are no tangible events (Aristotle 2007a). Going many steps further, Bergson defined time as a sequence of finite and heterogeneous durations. Time is irreversible, and the future unpredictable (Bergson 1910). Each duration has a significance different from that of each preceding and following one. The transition from inner time to the time of things is related to *memory* and *consciousness*. As Bergson (1965) explains, "Without an ultimate memory, which links two instants, the one to the other, there would be only the one or the other of the two, consequently a unique instant, without before and after, without succession, thus no time. . . . It is impossible to distinguish between the duration, however short it may be, which separates two instants and a memory that connects them, because duration is essentially a continuation of what no longer exists into what does exist. This is real time, perceived and lived. This is also any conceived time, because one cannot conceive a time without imagining it as perceived and lived. Thus, duration implies consciousness; and we place consciousness deeply in the things by crediting them a time that endures." Along the same lines Aristotle wonders, "Would there be time if there were no soul?" (Aristotle 2007b) and remarks, "When we are conscious of no mental change we do not think time has elapsed, any more than the fabled sleepers of Sardinia do when they awake; they connect the later now with the earlier and make them one" (Aristotle 2007a).

In contrast to deterministic clock time, Bergsonian time describes the temporal processes of irreversible dynamic systems. Deterministic models are unable to explain clock time's irreversibility, and thus unable to antici-pate the temporal complexity and dynamism of real systems. The Bergson-ian approach to time as duration is characterized by indeterminism, heterogeneity, and irreversibility, capturing the essence of human behav-ior. I would argue that Bergsonian time is the "time of social systems" and is built on the Einsteinian time of physical systems.

The new property that the Web contributes to the context of "social time" is that it enables the automatic and effortless recording of starting time and ending time for all visiting sessions. Durations are becoming not only discoverable and observable but also traceable, processable, and massive. These online visiting durations increase the material dimensions of networks.

Bergson's belief in the reversibility of time stems from the general irreversibility of a being. On the Web, the being is related to time in the sense that what it means for a Web being to be, is to exist temporally in the stretch between birth and death. While the birth of a Web being in physical time is the action of uploading and assigning an exclusive URI, birth in Web time comes from the first visitors. Analogously, death in physical time is the removal of Web being's URI and in Web time the total absence of visitors.

Antoniou and Christidis (2010) recently made the first step in formal-izing Bergsonian time by introducing the relevant time operator in terms of innovation and aging. Their approach is initiated by the definition of clock time as a scalar quantity, the values of which correspond to a subset of real numbers. The induced "age" of a random variable is the "age" of each time duration weighted by the probability of innovation, though Antoniou and Christidis have not incorporated the significance of dura-tions with consciousness and memory.

In general, the sustainability of existence on the Web is described as a cyclical economic function in inner Web space and time (see figure 5.1 in section 2.5.1). In this function, some businesses and governments (e.g., in China and Turkey) try to take control of Web space and time. Their usual approach is to model human behavior by reconstructing users' memory and consciousness based on their visiting sessions. For the moment, a small number of mammoth companies control most of Web space and time, restricting governments to passive and repressive policies (Vafopoulos 2012). On the basis of their control of vast amounts of collected and processed user data, Google, Facebook, Amazon, and Apple (to name only a few) act as gatekeepers of the online Web ecosystem. This fact undercuts the foundations of the Web and has started to raise serious complaints from the academic community. The "computational social science argument" (Lazer et al. 2009) can be summarized as the substantial barriers posed to scientists, mainly by business, to their collecting and analyzing the terabytes of available data, and describing the minute-by-minute interactions and locations of entire populations of individuals.

3. How the Web Affects Traditional Space, Time, and Being

3.1. Space and the Web

Let us now relax the hypothesis of the self-contained Web and describe how the Web affects physical space, time, and existence.

User logs of every visiting session register a user's IP address, which can be used to trace the ISP (Internet service provider) and computer terminal. In this way, Web space is directly and automatically connected to physical space. The Web space assists us in navigating the physical space, and vice versa. In this way the physical space becomes more *discoverable* and *traceable*. Many travelers find their destination through online maps, and farmers make decisions for their plants based on online weather forecasts. Google is now adjusting search results to take into account the declared place of origin of users for related queries. Generally, Web space tends to both expand and limit the notion of physical space. Hyperconnected users may carry their digital "Internet connection" anywhere, but it can also be the case that some of them may restrict their transportation due to the role of the Web as communication replacement (Wellman 2002). Besides massive navigation, aggregation, and recombination, the Web extends the *aspatiality* of existence. In economy, geography still matters, especially in the production of knowledge-intensive industries, whose synchronous face-to-face interactions and critical mass in human capital are important inputs. The major implication of aspatiality is that transportation and transaction costs are negligible, creating a new range of possibilities.

As Stiegler (2010) points out, "Post-globalization is not a territorial withdrawal: it is on the contrary the inscription of territory in a planetary reticularity through which it can be augmented with its partners at all the levels of which it is composed, from the interpersonal relation made possible by the opening up of rural regions implementing a politics of the digital age, to business which, deploying its competence locally and contributively, knows how to build a de-territorialized relational space: ecological relational space is a territory of hyper-learning."

3.2. Time and the Web

Before the advent of the Web, the notion of "market" was mainly used to describe physical meetings between sellers and buyers in order to exchange goods for a price in a specific time frame. Traditional marketplaces were based on synchronization and agreement over time, place, and price. In the Web market, participants can make decisions across a wider spectrum of choices. They cannot smell or touch but they can search for quality and price on an "around the clock" basis. Most human activities that take place through or on the Web have become available *asynchronously*, (*in part*) *synchronously*, and *continuously* (e.g., e-commerce, social networking).

If physical time is an arbitrary standard that enables the division of infinite space into useful parts, then the Web assists us in separating it into even finer pieces. What the Web imparts to physical time and space is *flexibility* and an enriched set of choices for human action. An increasing number of individuals, groups, and businesses reclaim this flexibility, putting pressure on traditional socio-economic structures, which are based on pre-Web physical time and space hypotheses.

3.3. Being and the Web

The resulting paradigm shift has circular effects on personal choice and collective action. In particular, the proliferation of wireless Internet and mobile devices has contributed to the transformation of human communities from densely knit "little boxes" (linking door to door) to sparsely knit "glocalized" networks (linking both locally and globally) and "networked individuals" (linking with little regard to space) (Wellman 2002).

Networked individuals mobilize part or all of their information and communication system, operating in a more *flexible*, *less-bounded*, and *spatially dispersed* environment. This innovative modus operandi includes frequent *switching* among multiple social networks and modes of communication, a different sense of belonging, flexible business arrangements, and intense time management. The emergence of the Web ecosystem has increased the quantity and quality of opportunities not only to discover and update information but also to communicate. The quantity could be simply approximated by the largest number of links among beings and Web beings. This impressive augmentation of physical space with Web space capacities has also affected the quality of human interaction. The Web broadens face-to-face communication by expanding the reach of social networks to multiple-channel communication. As Wellman (2002) explains, the Web "simultaneously affords: (a) personal communications between one or multiple friends, (b) within network broadcasts; and (c) public addresses to strangers."

The catalytic role of the Web in providing new opportunities for collaboration and creativity has been crystallized in the development of a new third realm, the *privatized space*. The privatized space arises between the private realm of intimacy and individualism and the public realm of citizenship and active participation for the societal good (including professional activity). The Web functions mainly as a privatized space, with public life, sociability, and public opinion, with public interactions and visibility, but with private reasoning and motivation (O'Hara and Shadbolt 2008).

In this new privatized space has emerged a decentralized peer production through loosely affiliated self-empowered entities as the third mode of production, a third mode of governance, and a third mode of property (Bauwens 2006). Peer-production communities are based on information-sharing mechanisms so as to create public knowledge repositories and encapsulate communities' aggregated preferences and expectations. Peer production constitutes a new form of decentralized intercreativity outside the traditional market and price mechanisms by redefining two economic orthodoxies: *diminishing marginal productivity* and *increasing returns to scale* (Vafopoulos 2012).

According to the law of diminishing marginal productivity, as inputs create less additional product at the margin, productivity of variable inputs declines as quantity increases. The law is valid for "lumpy" goods like cars and computers, which are characterized by fragmented divisibility in production tasks, specialized knowledge in multiple activities and high learning, coordination, and switching costs of production. Inputs could be indivisible because of cost, technology, regulation, or specialization limitations. Nevertheless, the opposite appears to be the case for a Wikipedia entry or a tweet, as many users make light work at the margin with low coordination costs. In the case of divisible inputs, unlimited numbers of users can contribute arbitrarily fine increments of input like sentences or photographs.

Peer communities are based on information-sharing mechanisms concerning inputs and/or outputs, which create public knowledge repositories for storing the aggregated preferences and expectations of their users. These collective memory mechanisms could be product reviews, tutorials, and guides, user reputation systems, collaborative filters, archived tags, and links or various types of metadata.

The lack of direct compensation and the temporal disconnection between effort and rewards are the shared characteristics in peer production, procurement, and patronage production models. However, peer production is strictly based on collaborative production mechanisms and supply-side knowledge externalities. Accordingly, peer production emerges as a new form of production that complements the existing form and extends David's taxonomy, adding a fourth *p* (property, procurement, patronage, and peer production) (David 1992).

As the distributed action enhances and partially substitutes groupware, the issues of security, privacy, identity, and trust on a massive scale become of primary importance.

3.4. Discussion

The Web transforms the way we work, shop, learn, communicate, search, participate, and recreate, introducing unexpectedly novel and complex actualities. Immaterial thoughts and preferences are becoming visible, describable, and traceable parts of collective existence. However, it is still difficult to fully understand and anticipate the cumulative effects of these extra choices on human societies. Engineering, business, and regulatory decisions about the Web often result in unanticipated outcomes. Questions about the viability, feasibility, and efficiency of the Web are high on many research agendas. Pessimists and optimists, technophobians and technophiles, engineers and social scientists, bloggers and CEOs question the present and the future of the Web with equal zeal.

What changes are required so that the Web can work better? The contribution of philosophical thinking in this interdisciplinary dialogue could be threefold: (a) to establish a common minimum understanding about existence, time, space, and the underlying values of the Web, (b) to identify the important issues arising from the Web, and (c) to build models and metrics that will enable decision making about the Web. This campaign will be successful if Web scholars from diverse disciplines incorporate into their research projects thoughts, arguments, and mechanisms related to the questions mentioned above.

I address point (a) here by defining being, space, and time on the Web. The Web "curves" physical time and space by adding flexibility, universality, more available options, and sources of risk. Its growing influence proliferates via the construction of detailed user profiles and related theories of behavior and collective action.

The emerging function of virtualization creates new possibilities and redistributes human choice, resulting in more complex relations. The plethora of (almost) continuous user data enables the joint analysis of entities and their attributes. In this context, Web scholars and regulators face two major strands of challenges:

(1) to obtain the right balance between open access to online information and self-determination of users, on the one hand, and to provide the proper incentives to produce content and develop network infrastructure, on the other, and

(2) to accelerate socio-economic development by facilitating life-critical functions in the developing world and by enabling transparency, participation, and added-value services in the developed world.

The aforementioned issues entail a series of second-level components that further elaborate their diverse facets. These components may refer to matters like the lack of credibility and quality of some Web beings, propaganda, defamation, rumors, addictogenic behaviors (Stiegler 2010), the waste of time of users, and so forth. The next steps in addressing this new form of complexity could be directed to describing the core functions of beings and Web beings in a computable framework, which builds on results in network and Web analysis.

I believe that the Web ecosystem is an ethically relevant social machine and should be systematically analyzed as such in order to realize its potential in promoting human values. In fact, the initial motivation behind the development of the Web itself was esteem, pride, excellence, absence of guilt, rewards, and indignation (O'Hara 2010). The ethical analysis of the Web as an integral part of philosophical studies should investigate how basic values like freedom are transformed through and on the Web.

As in physical space, ultimate value on the Web could be the ensuring of the free will of users to collectively transform Web beings (Wark 2004). The ongoing change in paradigm alters the content and means of achieving self-determination on and off the Web. Being self-determined on the Web is strongly related to underlying personal data management processes. These processes should assure anonymity in particular contexts, privacy, transparency, accountability (Weitzner et al. 2008), and auditing for both users and regulators. For the Web to be a free and open innovation platform, it should be engineered as net neutral (Vafopoulos 2012), not fragmented (e.g., Yeung et al. 2009) and uncensored (Kroes 2011) space. Public and private funding for independent institutions should sustain open and effective standards. Furthermore, it is of equal importance that regulatory policies acquire the right balance of market power and innovation, and favor Web-based development (Vafopoulos 2012 and 2005; Vafopoulos, Gravvanis, and Platis 2006).

The Web ecosystem is a unique opportunity to rethink and reengineer human life, and a prominent role of philosophy should be to communicate this challenge not only in the scientific and ethical tradition but also in the research sphere.

References

Abdulla, G. 1998. *Analysis and Modeling of World Wide Web Traffic*. Blacksburg, Va.: Virginia Tech.

Albert, R., and A. Barabási. 2002. "Statistical Mechanics of Complex Networks." *Reviews of Modern Physics* 74, no. 1:47–97.

Angwin, J. 2010. "The Web's New Gold Mine: Your Secrets." *Wall Street Journal*. At http://online.wsj.com/article/SB1000142405 2748703940904575395073512989404.html

Antoniou, I., and T Christidis. 2010. "Bergson's Time and the Time Operator." *Mind and Matter* 8, 2:185–202.

Aristotle. 2007a. *Physica IV 218b*. Translated by R. P. Hardie and R. K. Gaye. eBooks@Adelaide.

———. 2007b. *Physica IV 223a*. Translated by R. P. Hardie and R. K. Gaye. eBooks@Adelaide.

Bauwens, Michel. 2006. "The Political Economy of Peer Production." *Post-autistic Economics Review* 37:33–44.

Bergson, H. 1910. *Time and Free Will: An Essay on the Immediate Data of Consciousness*. Translated by F. L. Pogson. London: Sonnenschein.

———. 1922. "Société Française de Philosophie Conference." *Bulletin de la Société française de Philosophie*, 102–13.

———. 1965. *Duration and Simultaneity with Reference to Einstein's Theory*. Translated by Leon Jacobson. Indianapolis: Bobbs-Merrill.

Berners-Lee, T., D. Weitzner, W. Hall, K. O'Hara, N. Shadbolt, and J. Hendler. 2006. "A Framework for Web Science." *Foundations and Trends® in Web Science* 1, no. 1:1–130.

Castells, M. 2003. *The Power of Identity: The Information Age: Economy, Society and Culture, Volume II: The Information Age*. 2nd ed. Oxford: Wiley-Blackwell.

David, P.A. 1992. "Knowledge, Property, and the System Dynamics of Technological Change." *World Bank Research Observer* 7:215–15.

Goldfarb, A., and C. Tucker. 2010. "Privacy Regulation and Online Advertising." *Management Science* 57:57–71.

Halpin, H., and V. Presutti. 2009a. "An Ontology of Resources for Linked Data." *Linked Data on the Web (LDOW2009)*. At http://events.linkeddata.org/ldow2009/papers/ldow2009_paper19.pdf

———. 2009b. "An Ontology of Resources: Solving the Identity Crisis." In *The Semantic Web: Research and Applications*, 521–34.

Heidegger, M. 1962. *Being and Time*. Translated by J. Macquarrie and E. Robinson. New York: HarperCollins. (Original published in 1926.)

Hendler, J., N. Shadbolt, W. Hall, T. Berners-Lee, and D. Weitzner. 2008. "Web Science: An Interdisciplinary Approach to Understanding the World Wide Web." *Communications of the ACM* 51, no. 7:60–69.

Jackson, M. 2008. *Social and Economic Networks*. Princeton: Princeton University Press.

Kim, J. 2001. "Phenomenology of Digital-being." *Human Studies* 24, nos. 1–2 (March): 87–111. doi:10.1023/A:1010763028785

Kroes, N. 2011. "Online Privacy: Reinforcing Trust and Confidence." Online Tracking Protection & Browsers Workshop. At http://europa.eu/rapid/pressReleasesAction.do?reference=SPEECH/11/461&format=HTML&aged=0&language=EN&guiLanguage=en

Latour, B. 2002. "Gabriel Tarde and the End of the Social." In *The Social in Question: New Bearings in History and the Social Sciences*, ed. P. Joyce, 117–32. New York: Routledge.

———. 2005. *Reassembling the Social: An Introduction to Actor-Network Theory*. New York: Oxford University Press.

———. 2009. "Spheres and Networks: Two Ways to Reinterpret Globalization." *Harvard Design Magazine* 30 (Spring/Summer): 138–44.

———. 2011. "Networks, Societies, Spheres: Reflections of an Actor-Network Theorist." *International Journal of Communication* 5:796–810.

Lazer, D., A. Pentland, L. Adamic, S. Aral, A. Barabási, D. Brewer, N. Christakis, et al. 2009. "Social Science: Computational Social Science." *Science* 323, no. 5915 (February 6): 721–73.

Lévy, P. 1998. *Becoming Virtual: Reality in the Digital Age*. Translated by R. Bononno. New York: Plenum.

Monnin, A. 2009. "Artifactualization: Introducing a New Concept." In *InterFace 2009: 1st International Symposium for Humanities and Technology*. Vol. 3. Southampton, U.K.

———. 2011. "Philosophy of the Web as Artifactualization." In *Web-Science Montpellier Meetup*. At http://www.ppls.ed.ac.uk/ppig/documents/Philosophy%20of%20the%20Web%20as%20artifactualization.pdf

Negroponte, N. 1995. *Being Digital*. New York: Alfred A. Knopf.

———. 2000. "From Being Digital to Digital Beings." *IBM Systems Journal* 39, nos. 3–4 (July 2000): 417–18.

Newman, M. 2008. "The Mathematics of Networks." In *The New Palgrave Encyclopedia of Economics*, ed. L. Blume. London: Palgrave Macmillan. At http://www-personal.umich.edu/~mejn/papers/palgrave.pdf

O'Hara, K. 2010. "The Web as an Ethical Space." In *Proceedings of the British Library Workshop on Ethics and the World Wide Web, 16th November 2010, British Library, London*. London: British Library.

O'Hara, K., and N. Shadbolt. 2008. *The Spy in the Coffee Machine: The End of Privacy as We Know It*. Oxford: Oneworld.

Page, L., S. Brin, R. Motwani, and T. Winograd. 1998. "The PageRank Citation Ranking: Bringing Order to the Web, Technical Report." *Stanford InfoLab*. At http://ilpubs.stanford.edu:8090/422/

Pollock, R. 2005. "The 'New' Innovation Theory: Boldrin and Levine and Quah." http://www.rufuspollock.org/economics/papers/boldrin_levine_quah.html.

Quah, Danny. 2003. *Digital Goods and the New Economy*. London: Centre for Economic Performance.

Rivera-Batiz, L. A., and P. Romer. 1990. "Economic Integration and Endogenous Growth." *Quarterly Journal of Economics* 106, no. 2:531–55.

Ross, W. D. 1953. *Plato's Theory of Ideas*. Oxford: Clarendon Press.

Shapiro, C., and H. Varian. 1999. *Information Rules: A Strategic Guide to the Network Economy*. Cambridge, Mass.: Harvard Business Press.

Simon, H. A. 1971. "Designing Organizations for an Information-Rich World." *International Library of Critical Writings in Economics* 70:187–202.

Stiegler, B. 2010. "Manifesto." *Ars Industrialis*. At http://arsindustrialis.org/manifesto-2010

Vafopoulos, M. 2005. "A Roadmap to the GRID e-Workspace." *Advances in Web Intelligence, Lecture Notes in Computer Science* 3528:1054–56.

———. 2006. "Information Society: The Two Faces of Janus." *Artificial Intelligence Applications and Innovations* 204:643–48.

———. 2011a. "A Framework for Linked Data Business Models." In *15th Panhellenic Conference on Informatics (PCI)*, 95–99. At http://www.computer.org/portal/web/csdl/doi/10.1109/PCI.2011.74

———. 2011b. "Modeling the Web Economy: Web Users and Goods." In *WebSci '11, June 14–17, 2011, Koblenz, Germany*. The Web Science Trust.

———. 2011c. "Web Science Subject Categorization (WSSC)." In *Proceedings of the ACM WebSci'11*, 1–3. ACM. At http://journal.webscience.org/511/

———. 2012. "The Web Economy: Goods, Users, Models and Policies." *Foundations and Trends® in Web Science* 2, no. 5:384–461.

Vafopoulos, M., G. Gravvanis, and A. N. Platis. 2006. "The Personal Grid e-Workspace (g-Work)." In *Grid Technologies Transaction: State-of-the-art in Science and Engineering*, vol. 5, ed. H. Bekakos, M. Gravvanis, and G. Arabnia, 209–34. Billerica, Mass.: WIT Press.

Wark, M. K. 2004. *A Hacker Manifesto*. Cambridge, Mass.: Harvard University Press.

Watts, D. J. 2003. *Small Worlds: The Dynamics of Networks Between Order and Randomness*. Princeton: Princeton University Press.

Weitzner, D., H. Abelson, T. Berners-Lee, J. Feigenbaum, J. Hendler, and G. J. Sussman. 2008. "Information Accountability." *Communications of the ACM* 51, no. 6:82–87.

Wellman, B. 2002. "Little Boxes, Glocalization, and Networked Individualism." In *Digital Cities II: Computational and Sociological Approaches*, ed. M. Tanabe, P. van den Besselaar, and T. Ishida, 337–43. Berlin: Springer.

Yarow, J. 2011. "These Five Companies Control 64% of All Online Ad Spending." *Business Insider*. At http://www.businessinsider.com/chart-of-the-day-these-five-companies-control-64-of-all-online-ad-spending-2011-10#ixzz1f1FFBmPs

Yeung, C. A., I. Liccardi, K. Lu, O. Seneviratne, and T. Berners-Lee. 2009. "Decentralization: The Future of Online Social Networking." In *W3C Workshop on the Future of Social Networking Position Papers*. Vol. 2. At http://www.w3.org/2008/09/msnws/papers/decentralization.pdf

CHAPTER 7

EVALUATING GOOGLE AS AN EPISTEMIC TOOL

THOMAS W. SIMPSON

1. Knowledge and the Web

The Web raises tricky and important epistemological questions. Because the Web supports mass collaboration in a way hitherto unprecedented, one question is the epistemic value of projects such as Wikipedia (Fallis 2008, 2009; Tollefson 2009; de Laat 2010). Because the Web allows anyone to become an author, publishing to potentially vast audiences in what is also a hitherto unprecedented way, another question is whether and in what ways blogs are epistemically superior to offline media (Goldman 2008). My topic in this chapter is another issue: the epistemic value of search engines. This arises out of another hitherto unprecedented feature, the sheer quantity of online testimony. This means that, without help, you are unlikely to find what you want in a timely way. While the problem of retrieval caused by an information glut is not new, its scale and pervasiveness are. Novel ways of finding information have been developed to overcome this problem, with search engines such as Google, Bing, and Yahoo being the most successful current model. As a distinctive way of addressing the problem of information overload, search engines merit particular attention from social epistemologists. So I shall address three questions. First, what epistemic functions do search engines perform? Second, what dimensions of assessment are appropriate for the epistemic evaluation of search engines? Third, how well do current search engines perform on these?

The conclusions I shall reach are as follows. In answer to my first question, I argue that the epistemic role that search engines perform is that of a *surrogate expert*. They point the lay enquirer to valuable sources of information (section 2). In answer to my second question, I propose three dimensions of assessment as appropriate for the epistemic evaluation of a search engine. I label these *timeliness*, *authority promotion*, and *objectivity* (section 3). In answer to my third question, I argue that recent developments in how search engines rank their results impair achieving one of these dimensions, namely, objectivity (section 4).

"Personalisation" is the development in Internet-delivered services referred to, and consists in tailoring online content to what will interest the individual user. But objectivity may require telling enquirers what they do

Philosophical Engineering: Toward a Philosophy of the Web, First Edition. Edited by Harry Halpin and Alexandre Monnin. Chapters © 2014 The Authors except for Chapters 1, 2, 3, 12, and 13 (all © 2014 John Wiley & Sons, Ltd.). Book compilation © 2014 Blackwell Publishing Ltd and Metaphilosophy LLC. Published 2014 by Blackwell Publishing Ltd.

not want to hear, or are not immediately interested in. So personalisation threatens objectivity. Objectivity matters little when you know what you are looking for, but its lack is problematic when you do not. In an analogy, current search engines are powerful spotlights, but are little use for illuminating a wider area. There are two implications. First, traditional notions of epistemic authority are proving resilient on the Web. Some visions of the Internet see its value as consisting in the creation of non-hierarchical social space. But for enquirers who want objectivity, finding an expert in that area remains one of the best ways of achieving it—and this is a paradigm of hierarchy. Second, I propose an argument for the prima facie legitimacy of public intervention, as a means of promoting objectivity.

Before undertaking the analysis, and to forestall misunderstanding, it is important to note some limitations on the scope of this analysis.

For one, analysing the epistemic value of a practice does not imply that epistemological criteria of assessment are the *sole* or even the *primary* dimensions of evaluation. As this point holds generally, it applies in this specific case. Google and other search engines are rightly evaluated in terms of how well they respect users' privacy, for instance, which is an ethical concern. My starting point is the limited one, undoubtedly true: search engines *are* epistemically significant. Having noted this, the task I undertake is to evaluate them epistemically. Imperatives for change can only be conditional: *if* we value these epistemic outcomes, then the following changes are desirable. It is plausible that we do value these outcomes, and so the consequent enjoys similar support. But establishing the antecedent is a general practical and epistemological question and beyond my scope here.

For another, technology changes. Significant developments in the way search engines provide their results may well require changing the analysis, or even starting anew. All I claim is that given how search engines currently work, the dimensions of assessment I shall identify are important ways to assess their epistemic performance.

2. The Epistemic Role of Search Engines

Search engines are epistemically significant because they play the role of a surrogate expert. Take enquirers with the question, p or not-p? They can investigate whether p themselves. Or they can find someone who already knows whether p, and learn from that person. Call a person who already knows whether p, an *expert*. Frequently enough, consulting an expert will be a more reliable way of learning whether p than finding out yourself; I am not well able to assess the proof of Fermat's last theorem, for instance, and so defer to expert judgment. In addition, practical constraints on knowledge acquisition mean that, very often, finding an expert is quicker than finding it out yourself. This latter is significant for the terms in which an epistemic practice should be assessed. Although epistemic reasons and practical reasons are not the same (see Kelly 2003), a practice which is principally aimed at an *epistemic* outcome still ought to be, at least in part, assessed in *instrumental* terms. For it is a practical question how effective that practice is at yielding the desired result. Many issues in social epistemology are, in fact, practical questions about how well a process "commons" knowledge, to use Michael Welbourne's phrase (1986).

Expertise comes in degrees, and so we may distinguish two ends of a spectrum. On the shallowest notion of an expert, expertise with respect to p requires only that a person knows whether p. The eyewitness to the car crash, who saw the Volkswagen drive through the red lights, is an expert with respect to that particular event. There is also a deeper notion of an expert, as someone who has knowledge of both the intricacies and the big-picture issues in a particular domain through substantial investment of time and labour. A "deep" expert may not have exhaustive knowledge of his or her domain, but should have comprehensive knowledge. Expertise in this deeper sense is often institutionally acknowledged, through professional occupation, or membership of a guild, and so on. Significant additions to knowledge are typically attributable to these deep experts.

Deep experts perform functions that "shallow" ones cannot. Shallow experts merely offer testimony as to whether p, and, at the limit, are able to answer only the question whether p. Deep experts do more. Because of their comprehensive knowledge, deep experts are able to provide testimony which meets the diverse informational needs of different enquirers. As well as giving testimony, they can teach enquirers about how to understand their domain of expertise. This is because they can orient enquirers to what is important. They are able to do so both because of their comprehensive knowledge and their ability to make a judgment of relevance. In doing so, deep experts often refer enquirers to important sources. The expert's judgment on what sources are important is relied on by the enquirers in learning, until they are able accurately to make such judgments themselves. Think of what it is to teach undergraduates in philosophy. The range of philosophers that could be expounded or thinkers discussed is vast; for each great thinker or great movement, there are myriad under-labourers offering their commentary and development. The teacher's task is to pick out what is important. Teachers do more than this, of course—they also train enquirers in skills, for instance. When enquirers are new to a domain, and they lack the skills for understanding, or the domain is too large for them easily to gain comprehensive knowledge, they need a deep expert to make the relevant judgments that will orient them.

Now consider search engines. They perform two epistemic functions otherwise fulfilled by experts. One is a relatively simple one. Suppose an enquirer knows of some particular testimony online, but does not know exactly where it is; that is, he or she does not have the uniform resource identifier (URI) for the webpage or document with the testimony. Then a search engine can be used to find that testimony. The kind of expertise required here is a relatively limited one, and the function performed by a search engine here is not in principle different from that performed in the past by library card indexes. Andrei Broder (2002) terms this the "navigational" use of search engines.

Search engines also perform a richer function, traditionally one that deep experts were needed for. Current search engines do not themselves offer testimony in direct response to a question, in the way that IBM's Watson computer did on the TV quiz show *Jeopardy!* in February 2011.[1] The search engine Ask.com (also known as Ask Jeeves) is not a counter-example, for although its results have titles phrased as natural language answers to the presumed question behind the search, these are still links to other sources that the engine proposes as relevant. Nonetheless, search engines do perform the function of guiding the enquirer to sources of testimony which are supposed to be relevant to their information retrieval task, and which were previously unknown to the enquirer. The volume of content online means that it is now impossible for a single human to have comprehensive knowledge of the contents of the Web. But as much of the content of the Web is now indexed computationally, the processing power of computers means that the task can be accomplished automatically. With the appropriate algorithms, search engines can offer a judgment about likely relevant sources. So search engines now perform part of the function of a deep expert, in orienting the enquirer to relevant sources of information. Broder (2002) terms this the "informational" use of search engines.

When search engine results pages provide links to webpages and documents online, they imply that the target source is likely relevant to the enquirer's query. The results pages perform an additional function, however. By rank ordering the results, a further judgment is implied. When a link is ranked highly on the results page, it is implied that this is likely to be *more* relevant than those further down. In so far as one of the functions of a deep expert is to make a relevance judgment in orientating an enquirer to important sources of knowledge, not just relevant ones, search engines' decisions about which results to provide, in which rank order, is a quasi-fulfilment of the functions of a deep expert. Hence their epistemic function is that of a surrogate expert.

When discussing what the Pentagon knew about Saddam's Iraq prior to the 2003 invasion, U.S. Secretary of Defence Donald Rumsfeld famously divided the intelligence into "known knowns," "known

---

[1] For details of the research programme, see IBM 2012. As the applications are developed to process natural language more felicitously than at present, Watson-like abilities will no doubt be publicly available in future, delivered through the cloud in the way that search capabilities currently are.

unknowns," and "unknown unknowns."[2] The terminological difference between known unknowns and unknown unknowns makes the distinction sound more clear-cut than it actually is. In practice, as enquirers we have informational needs that are more or less precisely defined at the start of our enquiry, to the degree that sometimes we do not know what it is we are looking for but shall recognise it when we see it, while at other times we know exactly what it is that would satisfy the informational need. The categories of known unknowns and unknown unknowns felicitously describe the difference between the navigational and informational functions of search engines. Enquirers use search engines navigationally when they are looking for a known unknown. They use them informationally when they are looking for an unknown unknown, or at least something closer to that. The distinction is significant because the navigational and informational functions require different ways of assessing a search engine's epistemic utility, a problem to which I now turn.

3. Dimensions of Epistemic Assessment

How should search engines be epistemically assessed? In his seminal work on social epistemology, *Knowledge in a Social World* (1999), Alvin Goldman has noted the epistemic significance of search engines in a networked society as indispensible tools for coping with too much information. I first outline his framework and the two dimensions of assessment he proposes for epistemically evaluating search engines, before justifying and proposing three additional ones.

Goldman is interested in what he calls "veritistic value." True beliefs are intrinsically valuable, and truth-relevant belief states are assigned values to reflect this. True belief that *p* takes 1.0; not having a belief regarding *p* is the state of ignorance, and takes .5; believing *p* when not-*p* is the state of error and takes .0 (1999, 89). The *instrumental* veritistic value of a practice is then its propensity to increase or decrease the veritistic value of an individual or a population (1999, 87). In effect, a practice is approved if it leads to more people believing more truths and fewer falsehoods. I endorse Goldman's espousal of the value of truth, which he defends against "veriphobes," those who would raise a "suspicious and even scornful eyebrow at any serious attempt to wield the concept of truth" (1999, 7). (For further clarification of the project, see Goldman 2002. Longer defences of the value of truth include Blackburn 2005 and Lynch 2005.) At the least, falsehood is a critical objection to a belief.

---

[2] For a transcript, see Department of Defense 2002. Subsequent investigations revealed that the U.S. intelligence community collectively had sufficient evidence to justify the conclusion that Iraq possessed no weapons of mass destruction, but organisational failure meant that the conclusion had not been drawn. So the historical irony is that they didn't know what they knew, and they should have been looking for "unknown knowns"—the fourth category omitted by Rumsfeld's division.

Goldman then proposes two dimensions of assessment by which the instrumental veritistic value of a search engine should be measured. These are "its *precision* or its *recall*" (1999, 172). Precision is the ratio of relevant to irrelevant documents returned in response to a query. Recall is the ratio of total relevant documents returned in response to a query, to the total number of relevant documents on the Web.[3] Both precision and recall are ways of assessing the performance of a search engine on a single occasion. These can be generalised, however, giving a way to assess the overall performance of the search engine on each dimension of assessment. Assume a string of queries, $Q_1, \ldots, Q_n$. The search engine has a "precision" score for each. The average (mean) precision score for that string then gives a generalised score for that engine on precision. So too for recall. To some degree, it is arbitrary how the string of queries is defined, as there are infinitely many possible queries. The larger the string and more diverse the components, the more useful the score is as an indicator of general performance.

Clearly enough, precision and recall are both important ways of assessing the veritistic value of a search engine. Nonetheless, they do not exhaust the ways in which search engines should be epistemically assessed, and I shall argue that there are at least three further dimensions of assessment that must be added. Perhaps there are more, but my concern is only to establish that those I identify are also important. Because the navigational and informational uses of search engines are satisfied in different ways, there are different dimensions of assessment for each. So I address them in turn.

Although relatively easy to assess, the navigational function of search engines is not well measured by Goldman's two dimensions of assessment. Take the example of my search for an authoritative source for the quotation from Rumsfeld, in section 2 above, a navigational task. It was not clear in my memory whether it was Rumsfeld or Dick Cheney who said this, so my query string was: "know we know, know we don't know, don't know we don't know." Google returned "about 2,000,000,000 results." Of the first one hundred results, numbers 1, 11, and 59 were directly relevant to my enquiry, and numbers 5 and 76 were tangentially related. The precision of that search was not very high; making the optimistic supposition that the top one hundred results were a representative sample, a precision score of one in twenty is hardly encouraging. The recall score was unknowable, at least to me, because I could not determine the total number of relevant documents and webpages on the Web. In suggesting that this score may have been similarly poor, however, it is noteworthy

---

[3] Goldman's stated definition of recall is "the ratio of total relevant documents returned in a query to the total number of documents on the Web" (1999, 172), *not* "the total number of *relevant* documents on the Web," as I have it. As the ratio of relevant documents to total documents on the Web is not a figure that tells us anything about the success of the search engine, charity in interpretation requires the addition of the qualifier.

that the actual source I have deemed sufficiently authoritative—namely, the transcript of the press briefing on the Department of Defense website—was not returned in the top one hundred results. Nevertheless, despite the fact that Google's precision and recall were poor, the search was entirely successful. For the first result took me to an accurate quotation of Rumsfeld's words, which was exactly what I sought. So what is missing?

I propose that search engines should be assessed for what I term their *timeliness*. Timeliness is not the amount of time that it takes the search engine to return its results; this would be significant only if search engines were not as superbly quick as they are now. Rather, and roughly, timeliness is an indicator of the amount of time it is likely to take an enquirer to find a link that he or she wants on the results pages. More precisely, I specify that a search engine's timeliness in response to a particular query is given by the ranking of the first relevant link on the search engine's results pages; that is, the first link to a webpage or document which provides testimony that answers the enquirer's informational need. A generalised score for timeliness can be generated in the same way as for precision and recall, thereby assessing how well the search engine performs across many queries. (With one change: the average must be median, not mean, because of possible null results where a search engine fails to return a relevant link.)

In contrast to navigational uses of search engines, informational enquiries will typically require more than one relevant result. A variant of timeliness is clearly still important; it matters that relevant results be clustered towards the top of the ranking. So note a further criterion of assessment, which I term *distributed timeliness*. To specify this, assume that for any query Q_i there is a sufficient number of relevant results, where a relevant result is again a link to a webpage or document with testimony that answers the enquirer's informational need. The task is to evaluate the tightness of the clustering of the set of results which is sufficient for satisfying Q_i. This could be measured either by the set's average ranking (mean) or by its lowest-ranked result. Distributed timeliness is generalised in the same way as timeliness.

But epistemically valuable satisfaction of informational enquiries requires more than precision, recall, and distributed timeliness. I propose two further dimensions of assessment. Note that precision, recall, and timeliness are all dimensions of assessment that concern *relevant* documents and webpages. Relevance does not entail truth; testimony may be false or misleading but still purport to answer an enquirer's informational need. Online as much as offline, lies, confusion, hysteria, rumour, and what Harry Frankfurt terms "bullshit" are intermingled with sincere reports from knowledgeable people (Frankfurt 1988; Coady 2006 discusses other pathologies of testimony). I argue elsewhere that the fact that a speaker is a competent user of language does not justify an assumption that he or she is testifying sincerely (Simpson 2012). A search engine that prioritised links to webpages and documents on which there was *truthful* testimony, rather than merely *relevant* testimony, would be extremely epistemically valuable. Achieving this computationally is no doubt a challenging task. In practice, the search engine designers would need to find computable markers of (epistemic) authority. Institutional position would likely be one of the signals they could use. Hence my term for this criterion, *authority prioritisation*. How well a search engine prioritises epistemic authority is assessed in the same way as timeliness, but rather than links being "relevant," they must be "reliable," that is, be links to a webpage or document which provides truthful testimony that answers the enquirer's informational need.

While authority prioritisation would be extremely valuable, deep experts do more than point enquirers to reliable sources of information. So this is not yet to state how a surrogate expert should be assessed.

To do this, I term the final dimension of epistemic assessment that I propose, *objectivity*. Objectivity is important for enquirers using a search engine for an "informational" task. Here is a case that shows how precision and recall can be maximally fulfilled, yet a search engine still be criticisable. Suppose a searcher wishes to learn from eyewitnesses about atrocities committed during the recent overthrow of the Gaddafi regime in Libya, so he or she enters the following: "eyewitness AND reports AND libya AND atrocities AND 2011." For simplicity, presume that there are a hundred thousand reports on the topic online. Suppose that the search engine's results pages link to all and only these hundred thousand sites. Then both its precision and recall scores are 1, the maximum. Now suppose that half of these webpages and documents concern atrocities committed by Gaddafi loyalists, while the other half concern atrocities committed by National Transitional Council fighters, anti-Gaddafi militias, foreign forces, and all others. The same search engine results could return links to all those webpages and documents concerning atrocities committed by Gaddafi loyalists as the first fifty thousand hits, and relegate to the bottom fifty thousand links to those webpages concerning atrocities committed by anti-Gaddafi forces. The precision and recall scores would remain maximal. Yet this result would fail to satisfy an important epistemic norm. Because they are time-limited, searchers read only the small fraction of results at the top of the results pages. (One study found 79 percent of searchers did not go beyond the first results page, and 94 percent did not go beyond the second; Beitzel et al. 2007. For discussion and corroborating studies, see King 2008, 1–2.) Searchers who bothered to read only the results at the top of the results pages, and believed what they read, would come away with a seriously mistaken understanding of what happened. While they would have true beliefs about atrocities committed by one side, they would lack true beliefs about atrocities committed by the other. Yet an accurate understanding of the conflict requires this too. The nub of the problem is that precision and recall fail to recognise the significance of a link being placed top of page 1 of the results as against somewhere on page 16. The position of a link in the search results is an implied relevance judgment.

So there is a further criterion of epistemic assessment, in addition to precision, recall, and timeliness—namely, objectivity. A search engine's results are objective when their rank ordering represents a defensible judgment about the relative relevance of available online reports. The above case fails to be objective because rank ordering the results so that the top fifty thousand results link to reports of atrocities by one side only would not be a defensible judgment about the relative relevance of the available online reports. It is difficult to state what makes a judgment defensible without using terms that are either equivalent or themselves definable in terms of defensibility. But so long as we are able to recognise defensible judgments—and I take it we are—then this circularity is not problematic.

It might be thought that objectivity is entailed by successful authority prioritisation. And if so, then no attention need be paid to objectivity; focus on achieving successful authority prioritisation, and objectivity will follow. But objectivity is not so entailed. To complicate the example, suppose that of the fifty thousand reports of atrocities by Gaddafi loyalists, exactly half are truthful, and that the same is true of the fifty thousand reports of atrocities by opposition forces. Authority prioritisation would lead to all the truthful reports being prioritised above the false or misleading ones. But it would still be possible for the top twenty-five thousand results to link to testimony of Gaddafi forces' atrocities, and for the next twenty-five thousand results to link to testimony of opposition forces' atrocities. Such results would still fail to be objective. Because authority prioritisation is not sufficient for objectivity, objectivity must be assessed separately.

It may be objected that the Libyan atrocities case is a poor one and fails to show the search engine's result to be epistemically criticisable. This is because the content of the search query is ethically significant. We blame those who commit atrocities. So the searcher's resulting judgments of blame would be unfair, for they would blame only one side when both are guilty. The objection concludes that, because the impropriety of the search result can be explained in ethical terms, no epistemic impropriety need be supposed.

This objection fails. An example where the practical implications of the atrocities case are not present shows this. Consider a search for "important philosophers." In this case, nothing of practical significance hangs on your understanding of the history of philosophy; you are not preparing for an exam, for instance. Retain the basic structure of the atrocities case, but this time the top third of the results link to reports about German philosophers; the middle third link to reports about philosophers neither

German nor French (or plausibly both); the bottom third link to reports about French philosophers. If you trusted the search engine's implied relevance judgment but followed links from only the first few pages, you would come away with the view that there are no important French philosophers. This is plainly false. This search result would still be epistemically criticisable, and it would lack the dimension of ethical criticism present in the atrocities case. So objectivity is a genuinely epistemic value, as well as an at least sometimes ethical one.

Here is another possible objection. Goldman's criteria of precision and recall have the virtue of being quantifiable, whereas my objectivity criterion is not: for how can you calculate whether a judgment is defensible or not? So it is not a *useful* dimension of epistemic assessment, because no judgment about the relative merits of different search engines could be arrived at. I grant the objector's main claim, that a quantifiable judgment of objectivity cannot be made. But two points in reply. First, not all judgments must be quantifiable if they are to be useful. It is entirely routine for non-quantifiable judgments to be made that can nonetheless command widespread support. Quantifiable judgments have an element of epistemic coercion about them, allowing disagreement only on condition of irrationality, so it is argumentatively preferable to seek them when they are available. But their unavailability does not impugn the possibility of a useful assessment. Goldman also addresses this point: "My measures of [veritistic value] are intended to provide *conceptual* clarity, to specify what is *sought* in an intellectually good practice, even if it is difficult to determine which practices in fact score high on these measures. Conceptual clarity about desiderata is often a good thing, no matter what hurdles one confronts in determining when those desiderata are fulfilled" (1999, 91). Goldman then justifies the claim with an analogy of hiring someone for a job; clearly specifying the desired qualifications and properties of the successful candidate before advertising the position is a valuable task, even if working out which candidate best fulfils those is a tricky business.

As a second point in reply, it is important that judgments can be useful even if not quantifiable, for the objection falsely assumes that Goldman's criteria of precision and recall are quantifiable. They are not. This is because both rely on the notion of "relevance," and that is not a determinate criterion. If any concept is extensionally vague, "relevance" is. Because it may be vague whether a webpage is relevant, the precision and recall ratios could never be calculated. This is not problematic for Goldman, as noted, because the fact that neither the precision nor the recall criterion can be calculated does not mean they are not useful for epistemic assessment. The same point applies to timeliness, which also relies on the notion of relevance. Parallel points can be made about distributed timeliness and authority prioritisation, which rely on the notions of "sufficient" and "truthful," respectively. Both of these, again, are extensionally vague. But they are all capable of being recognised in clear cases.

The objectivity criterion is significant for a further reason: its value cannot be explained on Goldman's framework. Recall that his framework assesses epistemic practices according to their veritistic value, that is, according to whether they increase the number of true beliefs, eliminate ignorance, and correct error. This framework is too limited for a complete epistemic assessment. For one, it is unclear how to make sense of the implicit additive notion of accumulating true beliefs (see Treanor 2013). More importantly, though, we often want more than truth. Note that the original Libyan atrocities search result performs perfectly in terms of *veritistic* value, for all the testimony linked to would be true, and all the resulting belief states would take the value 1. Nonetheless, the result would still be criticisable. So there are other epistemic states we value in addition to true belief. Knowledge, for instance, is a state different from true belief, and while it is debated *why* we value knowing which is the road to Larissa, Plato was surely right to assume that it *is* preferable to having merely a true belief on the matter (*Meno* 97a–98a; for recent controversy on why knowledge is preferable, see Williamson 2000, 78–80; Hyman 2010; Millar 2011).

Nor is the notion of understanding explicable solely in terms of its veritistic value. It is possible to have knowledge of facts in a domain without having understanding of that domain; yet we value understanding over its lack. Someone understands a domain only if he or she has a substantial degree of objectivity, able to understand the domain independently of specific interests. My specified notion of search engine objectivity is intended to parallel this more general epistemic virtue, a virtue that only persons can possess.

In summary, then, I have proposed three ways of assessing the epistemic value of search engines—timeliness, authority prioritisation, and objectivity—in addition to the two that Goldman has proposed. These provide ways of assessing how felicitously a search engine would lead a time-limited enquirer to true beliefs. One of these dimensions of assessment, viz. objectivity, is challenged by recent developments in the design of search engines. I now turn to evaluate this challenge.

4. Personalisation and Objectivity

In this final section, I argue that the development of personalisation of search engines leads to worsening performance on my objectivity criterion. I then identify two implications of this conclusion. First, for the individual enquirer concerned with objectivity, I note the resilience of human judgments of relevance, as a way of bringing the enquirer's attention to important "unknown unknowns." Second, I claim that there are prima facie grounds for regulation of some sort. This is because there is a public good that is failing to be achieved.

Personalisation is a wider trend in how services are being provided over the Internet, beyond search engines alone. It consists in the use of algorithms to profile individual users, based on their past browsing and information consumption history, to predict what kinds of online content they will prefer. As profiles "deepen" over time, so the variance between what you and I are provided with when we go online increases. The idea is well captured by a quotation widely attributed to Mark Zuckerberg, founder of Facebook. "A squirrel dying in front of your house may be more relevant to your interests right now than people dying in Africa." If you really are more likely to follow a link that takes you to a squirrel dying in front of your house, personalised Web products accordingly prioritise that rather than the people dying in Africa. Personalisation on Facebook means that, in your news feed, it prioritises information about the friends you contact and follow the most. The Amazon recommender system will be known to nearly all present readers, and similarly works on the principle of personalisation: "People like you have also bought. . . ."

Pertinently, search engines now personalise your results. Entering exactly the same query, I may get a set of results different from the results you get. In practice, this means that sites you have previously visited will be prioritised in your search results pages; call this *individual personalisation*. In addition, sites that are visited by other people whose browsing histories resemble yours in ways picked out by the algorithms as relevant are also prioritised; call this *profile personalisation*. Google rolled out personalised search to signed-in users (those with Gmail or Google accounts) during the summer of 2005. In late 2009, it announced that personalized search would be the default option for all users, and this is the current situation.[4] Although you retain the power to turn the feature off, "the devil is in the defaults" (Kerr 2010). The significant majority of users are either unaware of changes to defaults, or do not know how to change them if they are, or do not care. Although we do not know the numbers—Google guards its data closely—the legitimate presumption is that the great majority of searches now are personalised.

Personalised search is a natural development for the company. Google co-founder Larry Page described "the 'perfect search engine' as something that 'understands exactly what you mean and gives you back exactly what you want'" (Google 2011). Personalisation is important in the specific context of search engines because it helps overcome problems of ambiguity in understanding intent. *My* search on "SEP" may be for the *Stanford Encyclopedia of Philosophy*; *yours* may be for the Severn Estuary

---

[4] See Google 2005, 2009. The differences between "signed-in" and "signed-out" personalised search are explained at Google 2012.

Partnership. Your history of browsing green-energy-related sites and mine of browsing philosophy-related sites provides the clue to the search engine to disambiguate. Personalised search is not restricted to Google, the undoubted market leader; Bing is also developing the capability.[5]

How does personalisation threaten objectivity? It is a contingent psychological fact about humans that we suffer from confirmation bias. That is, people are generally more likely to find reasons to discount evidence that goes against their antecedent belief and not to subject to similar scrutiny evidence that confirms their existing belief. They are also likely to view evidence that is consistent with their belief as confirmatory, even though it is also consistent with competing possibilities. Francis Bacon observed the problem: "Once a man's understanding has settled on something (either because it is an accepted belief or because it pleases him), it draws everything else to support and agree with it" (2000, § 46). The claim is empirically demonstrated. (For classic studies, see Wason 1960; Lord, Ross, and Lepper 1979. For an overview, see Nickerson 1998.) Confirmation bias means that we are, generally, more likely to seek out testimony that supports our existing beliefs, rather than testimony that contradicts it.

The problem is that personalisation reinforces confirmation bias. Suppose on past queries you have followed links only or predominantly to testimony that the snapshot suggests you will agree with. Then links to these sites are promoted in future searches, which, broadly, will offer testimony you are likely to agree with. So pages that contradict or challenge your existing views are likely to be demoted in the rankings. I am not the first to observe this. Eli Pariser (2011) recently coined the phrase "the filter bubble" to describe the effect of personalisation across the Web, including search engines, and he gives a number of (admittedly anecdotal) examples. Cass Sunstein (2007, 2008) observes a related phenomenon in the blogosphere, one that is self-imposed rather than algorithmically imposed. So personalisation threatens objectivity. The more personalised the results, the less they represent the sides of the argument you disagree with, and so the less objective they are.

There is a difficulty in extending the analysis beyond this general inverse correlation between personalisation and objectivity. The major search engines do not publicly declare all of the different factors they use to rank sites in response to a query, nor how heavily different signals are weighted. This is for obvious commercial reasons, both to retain competitiveness and because public disclosure would allow the marketeers to "game" the algorithms and push their favoured site higher than it merits on results pages. For those not on the inside, the degree to which search results are personalised is unknowable. But it is only *the degree* to which

---

[5] Google's market share of global search engine traffic is reported at 78 percent; see NetMarketShare 2012. For Bing's release of "adaptive search," see Bing 2011.

personalisation challenges objectivity that is unknowable; the fact that it does is shown by the argument above.

When the bubble is based on individual personalisation, only "epistemic saints" are immune from its effects—the people who scrupulously ensure that they seek out opposing views for every search they conduct in which there are importantly opposed views on the matter. The rest of us, one hopes, make the effort to seek out opposing views in serious enquiries. But the bubble is reinstated by our lazier browsing habits, when we prefer reading the testimony of those who broadly agree with us. In reflecting back to you results that confirm your antecedent belief, personalised search engine results constitute a *less* defensible judgment about the relative relevance of available online reports than do non-personalised results. When it comes to profile personalisation, even epistemic saints are in the same situation as the rest of us. For the results they receive come to reflect in part others' confirmation bias. The result of this is that your chance of having "unknown unknowns" promoted to you diminishes with personalisation.

Bacon, again, observed the causal effects of happenstance on people, in so far as we naturally form beliefs for reasons significantly to do with upbringing, authority, and the company we keep: "The evident consequence is that the human spirit (in its different dispositions in different men) is a variable thing, quite irregular, almost haphazard. Heraclitus well said that men seek knowledge in lesser, private worlds, not in the great or common world" (2000, § 42). Recalling Plato's allegory, Bacon termed the varying tendencies we have to form beliefs for reasons unconnected to truth the "idols of the cave."[6] My contention is that personalised search *supports* the idols of the cave, rather than helping to remove them. It is no doubt an unintended consequence, but it is a consequence nonetheless.

What are the implications of personalised search? I highlight two. The first concerns the response of the rational enquirer, and may be briefly stated. While many are unconcerned about objectivity in enquiry and thus are unconcerned about personalisation, many *are* concerned. For those who are, a way to avoid the filter bubble is to change the settings to turn off personalisation. Another option is to use a search engine such as DuckDuckGo, which returns non-personalised results as a by-product of its privacy policy of storing no user data (see http://duckduckgo.com). In the case of non-technologically aware enquirers, however, it is perhaps most likely that they will do what they previously did in the offline world: find the person who is reliable on the topic they want information about. Because of the quantity of online testimony, no single person can provide the one-stop shop. But collaborative projects can aim to do so, either generally, as in Wikipedia, or in more restricted domains. In *What Computers Still Can't Do* (1992), Hubert Dreyfus argues at length that

---

[6] Retaining the traditional but inaccurate translation of *idola*, "illusions."

practical judgments of relative importance are something that people are good at, because embodied and therefore skilled, and artificial reasoning systems such as algorithms are not. Regardless of whether the argument is sound, his conclusion remains true. Defensible relevance judgments are something that people are good at, and algorithms not. At least, they are not good at it yet.

The second implication is more substantial. There is an argument for the prima facie legitimacy of regulation of search engine providers. This is because there is a public good that is failing to be achieved. In 2005, the Pew Internet and American Life Project reported on the rise of search engines, and surveyed users' knowledge of how they worked. It concluded that "search engines are attaining the status of other institutions—legal, medical, educational, governmental, journalistic—whose performance the public judges by unusually high standards, because the public is unusually reliant on them for principled performance" (Pew 2005, 27). This is certainly correct (Halavais 2008 explores their social effects further). Given this, I make the case for intervention when the functioning of the marketplace has collectively damaging effects. It is appropriate that businesses should, in economic terms, carry the costs of their "externalities." Because externalities are here incalculable, and so no post hoc compensation is possible, the regulatory requirement ought to be on how the service functions in the first place.

For all the criteria of assessment except objectivity, there is an alignment between the interests of search engine operators, the preferences of the private individuals who use the service, and publicly desirable outcomes. Search engines' core business models are structured around advertising; Google provides a free service to enquirers, making money by providing sponsored links. Each time an enquirer clicks on a sponsored link, a small amount of income is generated for Google. The higher the number of enquirers who click on sponsored links, the higher Google's revenue. So it is in Google's interest to provide as excellent a service as possible to the enquirer, to maximise the number of enquirers who use the search engine. Sheer volume of traffic is the strategy. Given that precision, recall, timeliness, generalised timeliness, and authority promotion are all dimensions of search engine performance that enquirers desire, it is in Google's interest to perform well on these. There is no reason to suppose that these outcomes are anything but publicly desirable.

Objectivity is different. Other things equal, search engine operators have an interest in providing a service that supports the preferences of individual users for the commercial reasons outlined. Descriptively—and pessimistically—most people do not care terribly much about objectivity. So there is no commercial incentive for search engine operators to provide a service that does well by the objectivity criterion. But there *is* reason to think that a lack of objectivity is publicly undesirable. Given this, there is a divergence between the preferences of individuals and the interests of search engine operators on the one hand and public good on the other. Achieving a public good that would not otherwise be achieved is a classic prima facie justification for intervention by government. So there is a prima facie case for justified intervention by government in how search engines work.

Arguing for the permissibility of intervention sounds dramatic. In practice, a light hand can achieve significant results. As already mentioned, the devil is in the defaults. Simply requiring a change of the defaults from personalised to non-personalised search may be sufficient to achieve this public aim.

Why is objectivity a public good? This is the crucial premise of the proposal. I claimed above (section 3) that objectivity is necessary for understanding, and it is understanding that is important. No doubt there is more than one justification, but I shall focus on one that derives from "epistemic democracy." Epistemic democrats claim that democratic voting procedures are more likely than any other governance system to arrive at the outcome that is the best available for the voting population. This is not because of any legitimacy-bestowing properties that individual participation has (although epistemic democrats need not deny this); it rests on the thinner claim that voting is an epistemically exemplary way to learn what the best outcome is (Grofman and Feld 1988; List and Goodin 2001; Estlund 2009). Condorcet's Jury Theorem is a mathematical result and provides the basic framework. It states that the probability of a group of electors choosing the best available result approaches certainty as the number of electors increases. This is conditional, pertinently, on the proposition that individual electors on average perform better than chance in choosing the best available result. While electors' votes depend on a judgment about which option is the best available, that judgment can be acquired only through some degree of understanding of the issues at stake. So epistemic democracy requires some degree of objectivity on the part of the electorate—that degree of objectivity required for sufficient understanding of the issues involved such that they do better than chance in choosing the best available option. Epistemic democracy provides a justification for taking understanding to be valuable, and *a public good*. It is collectively valuable for individuals to possess understanding on political matters, and therefore for them to form views with objectivity. Search engines are now so socially important, because they are used so routinely as epistemic tools for citizens to form politically significant views, that they ought to return objective results.

It may be objected that epistemic democracy justifies a requirement for objectivity only on issues that are politically important. I grant the point. Nonetheless, *a lot* of issues are politically important. Economics, history, personal reputation, religion, morals, and science are all matters that are politically significant, so objectivity on these is required. There is, further, no sharp dividing line between issues that are politically important and those that are not. So search engine operators' responsibility for objectivity, justified by the good of epistemic democracy, is extensive, and the operators may legitimately be required by government to ensure that their search engines achieve some degree of objectivity over a wide range of matters.

Acknowledgments

I am grateful to Alex Oliver and two anonymous referees for comments and criticism on an earlier draft. I have also benefitted from conversation on these matters with John Naughton and Francis Rowland, and from comments from audiences at the Trust and Cloud Computing Conference and the Arcadia Project Seminar, both in Cambridge. I am grateful to Microsoft Research Cambridge, which supported research for this chapter through a Ph.D. studentship. The views expressed are my own and do not necessarily represent those of Microsoft Research.

References

Bacon, Francis. 2000 [1620]. *The New Organon*. Edited by Lisa Jardine and Michael Silverthorne. Cambridge: Cambridge University Press.

Beitzel, Steven M., Eric C. Jensen, Abdur Chowdhury, Ophir Frieder, and David Grossman. 2007. "Temporal Analysis of a Very Large Topically Categorized Web Query Log." *Journal of the American Society for Information Science and Technology* 58, no. 2:166–78.

Bing. 2011. "Adapting Search to You." At http://www.bing.com/community/site_blogs/b/search/archive/2011/09/14/adapting-search-to-you.aspx, retrieved 23 March 2012.

Blackburn, Simon. 2005. *Truth: A Guide*. Oxford: Oxford University Press.

Broder, Andrei. 2002. "A Taxonomy of Web Search." *ACM SIGIR Forum* 36, no. 2:3–10.

Coady, C. A. J. 2006. "Pathologies of Testimony." In *The Epistemology of Testimony*, edited by Jennifer Lackey and Ernest Sosa, 253–71. Oxford: Oxford University Press.

De Laat, Paul B. 2010. "How Can Contributors to Open-Source Communities Be Trusted? On the Assumption, Inference, and Substitution of Trust." *Ethics and Information Technology* 12, no. 4:327–41.

Department of Defense. 2002. "DoD News Briefing—Secretary Rumsfeld and Gen. Myers." At http://www.defense.gov/Transcripts/Transcript.aspx?TranscriptID=2636, retrieved 23 March 2012.

Dreyfus, Hubert L. 1992. *What Computers Still Can't Do: A Critique of Artificial Reason*. Cambridge, Mass.: MIT Press.

Estlund, David M. 2009. *Democratic Authority: A Philosophical Framework*. Princeton: Princeton University Press.

Fallis, Don. 2008. "Toward an Epistemology of Wikipedia." *Journal of the American Society for Information Science and Technology* 59, no. 10:1662–74.

———. 2009. "Introduction: The Epistemology of Mass Collaboration." *Episteme* 6, no. 1:1–7.

Frankfurt, Harry. 1988. "On Bullshit." In his *The Importance of What We Care About: Philosophical Essays*, 117–33. Cambridge: Cambridge University Press.

Goldman, Alvin. 1999. *Knowledge in a Social World.* Oxford: Oxford University Press.

———. 2002. "What Is Social Epistemology? A Smorgasbord of Projects." In his *Pathways to Knowledge: Private and Public*, 182–204. Oxford: Oxford University Press.

———. 2008. "The Social Epistemology of Blogging." In *Information Technology and Moral Philosophy*, edited by Jeroen Van Den Hoven and John Weckert, 111–22. Cambridge: Cambridge University Press.

Google. 2005. "Search Gets Personal." At http://googleblog.blogspot.com/2005/06/search-gets-personal.html, retrieved 23 March 2012.

———. 2009. "Personalized Search for Everyone." At http://googleblog.blogspot.com/2009/12/personalized-search-for-everyone.html, retrieved 23 March 2012.

———. 2011. "Technology Overview." At http://www.google.com/corporate/tech.html, retrieved 23 March 2012.

———. 2012. "Basics: Search History Personalization." At http://www.google.com/support/accounts/bin/answer.py?answer=54041, retrieved 23 March 2012.

Grofman, Bernard, and Scott L. Feld. 1988. "Rousseau's General Will: A Condorcetian Perspective." *American Political Science Review* 82, no. 2:567–76.

Halavais, Alexander. 2008. *Search Engine Society.* Cambridge: Polity Press.

Hyman, John. 2010. "The Road to Larissa." *Ratio* 23, no. 4:393–414.

IBM. 2012. "IBM/WATSON." At http://www-03.ibm.com/innovation/us/watson/, retrieved 23 March 2012.

Kelly, Thomas. 2003. "Epistemic Rationality as Instrumental Rationality: A Critique." *Philosophy and Phenomenological Research* 66, no. 3:612–40.

Kerr, Ian. 2010. "The Devil Is in the Defaults." *Ottawa Citizen* (29 May). Available at http://iankerr.ca/wp-content/uploads/2011/08/The-devil-is-in-the-defaults.pdf, retrieved 23 March 2012.

King, Andrew. 2008. *Website Optimization: Speed, Search Engine and Conversion Rate Secrets.* Sebastopol, Calif.: O'Reilly Media.

List, Christian, and Robert Goodin. 2001. "Epistemic Democracy: Generalizing the Condorcet Theorem." *Journal of Political Philosophy* 9, no. 3:277–306.

Lord, Charles G., Lee Ross, and Mark R. Lepper. 1979. "Biased Assimi-
lation and Attitude Polarization: The Effects of Prior Theories on
Subsequently Considered Evidence." *Journal of Personality and Social
Psychology* 37, no. 11:2098–109.
Lynch, Michael P. 2005. *True to Life: Why Truth Matters.* Cambridge,
Mass.: MIT Press.
Millar, Alan. 2011. "Why Knowledge Matters." *Proceedings of the Aris-
totelian Society* supplementary volume 85, no. 1:63–81.
NetMarketShare. 2012. "Search Engine Market Share." At http://
marketshare.hitslink.com/search-engine-market-share.aspx?qprid=4,
retrieved 23 March 2012.
Nickerson, Raymond S. 1998. "Confirmation Bias: A Ubiquitous Phe-
nomenon in Many Guises." *Review of General Psychology* 2, no. 2:175–
220.
Pariser, Eli. 2011. *The Filter Bubble: What the Internet Is Hiding from You.*
London: Penguin.
Pew Internet and American Life Project. 2005. "Search Engine Users."
At http://www.pewinternet.org/~/media//Files/Reports/2005/PIP_
Searchengine_users.pdf.pdf, retrieved 23 March 2012.
Simpson, Thomas W. 2012. "Testimony and Sincerity." *Ratio* 25, no.
1:79–92.
Sunstein, Cass. 2007. *Republic.com 2.0.* Princeton: Princeton University
Press.
———. 2008. "Democracy and the Internet." In *Information Technology
and Moral Philosophy*, edited by Jeroen Van Den Hoven and John
Weckert, 93–110. Cambridge: Cambridge University Press.
Tollefson, Deborah Peron. 2009. "*WIKIPEDIA* and the Epistemology of
Testimony." *Episteme* 6, no. 1:8–24.
Treanor, Nick. 2013. "The Measure of Knowledge." *Noûs* 47, no. 3:577–
601.
Wason, P. C. 1960. "On the Failure to Eliminate Hypotheses in a Con-
ceptual task." *Quarterly Journal of Experimental Psychology* 12, no.
3:129–40.
Welbourne, Michael. 1986. *The Community of Knowledge.* Aberdeen:
Aberdeen University Press.
Williamson, Timothy. 2000. *Knowledge and Its Limits.* Oxford: Oxford
University Press.

CHAPTER 8

THE WEB-EXTENDED MIND

PAUL R. SMART

Introduction

In the few decades since its invention the World Wide Web has exerted a profound influence on practically every sphere of human activity. Online stores have transformed the way we purchase goods, social networking sites have transformed the way we stay in touch with friends, and real-time news feeds have transformed the way we stay abreast of current affairs. For better or worse, it seems, the Web is poised to have a significant influence on the way we live our lives, and perhaps ultimately it will come to influence the social, political, and economic forces that determine how our lives are lived.

The rapid growth and penetration of the Web raises important questions about its effects, not just on our social activities, but also on the nature of our cognitive and epistemic profiles. The Web is a transformative technology, but its transformative influence does not necessarily stop at the social processes that govern our everyday interactions with one another. Many technologies that have transformed society (for example, the clock, the map, and systems of writing) have also exerted subtle (and sometimes not so subtle) effects on our cognitive and intellectual capabilities. The Web provides new opportunities for interaction and engagement with a global space of information, and, in some cases, such interactive opportunities may contribute to profound shifts in the way we see ourselves and the nature of our cognitive processing. But what exactly is the nature of this transformative influence, and by what mechanisms does the Web stand poised to transform our cognitive and epistemic capabilities?

One answer to this question comes in the form of the extended mind hypothesis (Clark and Chalmers 1998), which has been the focus of much recent philosophical debate (Adams and Aizawa 2008; Clark 2008; Menary 2010; Rowlands 2010; Rupert 2009; Smart 2010; Smart et al. 2010). The basic idea of the extended mind is that human cognition is sometimes constituted by information-processing loops that extend beyond the biological brain. It maintains that the machinery of human cognition can sometimes extend beyond the neural realm to include elements of our social and technological environments. When applied to the specific sociotechnical context of the Web, the extended mind hypothesis gives us the

Philosophical Engineering: Toward a Philosophy of the Web, First Edition. Edited by Harry Halpin and Alexandre Monnin. Chapters © 2014 The Authors except for Chapters 1, 2, 3, 12, and 13 (all © 2014 John Wiley & Sons, Ltd.). Book compilation © 2014 Blackwell Publishing Ltd and Metaphilosophy LLC. Published 2014 by Blackwell Publishing Ltd.

notion of the "Web-extended mind," or the idea that the technological and informational elements of the Web can (at least sometimes) serve as part of the mechanistic substrate that realizes human mental states and processes (Halpin, Clark, and Wheeler 2010; Smart et al. 2010; Smart et al. 2009).

This chapter attempts to explore the notion of the Web-extended mind. It first provides an overview of cognitive extension and the extended mind hypothesis, and it then goes on to discuss the possibility of Web-based forms of cognitive extension involving current or near-future technologies. It is argued that while current forms of the Web may not be particularly suited to the realization of Web-extended minds, new forms of user interaction technology as well as new approaches to information representation on the Web do provide promising new opportunities for Web-based forms of cognitive extension.

Cognitive Extension and the Extended Mind

The traditional view in the sciences of the mind sees the human brain as occupying a rather special place in the material fabric associated with the realization of human mental states and processes. The view claims that the machinery of the mind is housed largely within the head, and that to understand more about our cognitive profiles we need to understand more about the workings of the biological brain. Eventually, it is claimed, we will have a complete theory of human cognition, and within this theory the human brain will occupy centre stage.

The validity of this neuro-centric, or intra-cranial, perspective has recently been challenged by those who embrace situated, embodied, or distributed approaches to cognition (Hutchins 1995; Robbins and Ayded 2009; Clark 1999; Haugeland 1998; Pfeifer and Bongard 2007). Such approaches challenge the notion that mind and cognition are solely internal (neural) phenomena by emphasizing the role played by extra-neural and extra-bodily factors in shaping the profile of much real-world cognitive processing. One view that is perhaps maximally opposed to the traditional view of the human mind is the thesis of the extended mind (Clark and Chalmers 1998). This view actually goes by a variety of names, including locational externalism (Wilson 2000, 2004), active externalism (Clark and Chalmers 1998), vehicle externalism (Hurley 1998; Rowlands 2006), environmentalism (Rowlands 1999), cognitive extension (Clark 2008), and the extended mind (Clark and Chalmers 1998); however, what all of these locutions have in common is a commitment to the idea that aspects of the external, extra-neural environment can play a constitutive role in the material realization of human mental states and processes.

As an example of extended mind theorizing, consider the case of multiplying two three-digit numbers. One account of how we are able to multiply the two numbers might emphasize how we first derive some symbolic encoding of the visual (or auditory) input corresponding to the two numbers. It would then invoke a computational account according to which the inner symbols are manipulated in some way so as to achieve the correct mathematical outcome. Now contrast this with what is surely a more accurate (and ecologically realistic) picture of how we implement long multiplication in the real world. This alternative picture involves the active manipulation of external symbols in such a way that the problem confronting the biological brain is profoundly simplified. In place of purely inner, environmentally decoupled, computational operations we see a pattern of real-world-involving perception-action cycles—ones in which single-digit numbers are compared and intermediate computational results are stored in an external medium using (for example) pen and paper. This example, described in Wilson and Clark (2009), is a case of what we might call environmentally extended cognition or cognitive extension (see also Clark 2008). It takes what is ostensibly an inner cognitive capability (an ability to do long multiplication) and shows how crucial aspects of the problem-solving process can be (and usually are) delegated to aspects of the external world.

In their original formulation of the extended mind hypothesis, Clark and Chalmers (1998) attempted to go beyond the simple case of cognitive extension, as exemplified by long multiplication. In particular, they wished to show that extended mind theorizing could also be applied to cases involving the ascription of intentional mental states, such as states of belief and desire. In order to make this case, Clark and Chalmers (1998) asked us to imagine two individuals: Inga and Otto, both of whom are situated in New York City. Inga is a normal human agent with all the usual cognitive competences, but Otto suffers from a mild form of dementia and is thus impaired when it comes to certain acts of information storage and recall. To attenuate the impact of his impairment on his daily behaviour, Otto relies on a conventional notebook, which he uses to store important pieces of information. Otto is so reliant on the notebook and so accustomed to using it that he carries the notebook with him wherever he goes and accesses it fluently and automatically whenever he needs to do so. Having thus set the stage, Clark and Chalmers (1998) ask us to imagine a case where both Otto and Inga wish to visit the Museum of Modern Art to see a particular exhibition. Inga thinks for a moment, recalls that the museum is on Fifty-third Street and then walks to the museum. It is clear that in making this episode of behaviour intelligible (or psychologically transparent) to us Inga must have *desired* to enter the museum, and it is clear that she walked to Fifty-third Street because she *believed* that that was where the museum was located. Obviously, Inga did not believe that the museum was on Fifty-third Street in an occurrent sense (i.e., she has not spent her entire life consciously thinking about the museum's location); rather, she entertained the belief in a dispositional sense. Inga's belief, like perhaps many of her beliefs, was sitting in memory, waiting to be accessed as and when needed.

Now consider the case of Otto. Otto hears about the exhibition, decides to visit the museum, and then consults his notebook to retrieve the museum's location. The notebook says the museum is on Fifty-third Street, and so that is where Otto goes. Now, in accounting for Otto's actions we conclude, pretty much as we did for Inga, that Otto *desired* to go to the museum and that he walked to Fifty-third Street because that is where he *believed* the museum was located. Obviously, Otto did not believe that the museum was on Fifty-third Street in an occurrent sense (Otto has not spent much of his life constantly looking at the particular page in his notebook containing museum-related facts); rather, he entertained the belief in a dispositional sense. Otto's belief, like perhaps many of his beliefs, was sitting in the notebook, waiting to be accessed as and when needed.

Clark and Chalmers (1998) thus argue that the case of Otto establishes the case for a form of externalism about Otto's states of dispositional believing. The notebook, they argue, plays a role that is functionally akin to the role played by Inga's onboard bio-memory. If this is indeed the case, then it seems to make sense to see the notebook as part of the material supervenience base for some of Otto's mental states, specifically his states of dispositional belief (such as those involving museum locations). The main point of the argument is to establish a (potential) role for external artefacts in constituting the physical machinery of at least some of our mental states and processes. If, as Clark and Chalmers (1998) argue, the functional contribution of an external device is the same as that provided by some inner resource, then it seems unreasonable to restrict the material mechanisms of the mind to the inner, neural realm. It seems possible, at least in principle, for the human mind to occasionally extend beyond the head and into the external world.

Such claims are, understandably, disconcerting, and it is important that we understand the precise nature of the claim that is being made. One immediate cause for concern relates to the notion of functional equivalence between the inner (e.g., bio-memory) and outer (e.g., notebook) contributions. If we allow any form of externally derived influence to count as part of the mechanistic substrate of the mind, then doesn't this cast the mechanistic net too widely? Don't we end up confronting cases that are so blatantly counter-intuitive that they undermine the very notion of the mind as a proper locus of scientific and philosophical enquiry? Clearly, not all of the technologies or external resources that we encounter are apt to engage in the kind of bio-technological hybridization envisioned by the extended mind hypothesis. As Clark argues: "There would be little value in an analysis that credited me with knowing all the facts in the *Encyclopaedia Britannica* just because I paid the monthly installments and found space for it in my garage" (1997, 217).

Similarly, it would be foolish to equate my personal body of knowledge and beliefs as co-extensive with the informational contents of the Web simply because I own an iPhone. What, then, are the conditions under which we count a set of external resources as constituting part of an environmentally extended mind? In answering this question, Clark and Chalmers (1998) embrace a particular set of criteria, ones that appeal to the accessibility, portability, reliability, and trustworthiness of the external resource. The criteria are that:

1. "the resource be reliably available and typically invoked" (Clark 2010, 46) (Availability Criterion);
2. "any information thus retrieved from [the non-biological resource] be more or less automatically endorsed. It should not usually be subject to critical scrutiny (unlike the opinions of other people, for example). It should be deemed about as trustworthy as something retrieved clearly from biological memory" (Clark 2010, 46) (Trust Criterion); and
3. "information contained in the resource should be easily accessible as and when required" (Clark 2010, 46) (Accessibility Criterion).

Clearly, such criteria serve to guide and constrain our intuitions about the kind of bio-artifactual and bio-technological couplings that are relevant to the formation of an extended mind. And they do so precisely because they delimit the range of situations under which we recognize the capabilities engendered by an external resource as being (most plausibly) those of a specific individual human agent (or perhaps a collection thereof [Tollefsen 2006; Theiner, Allen, and Goldstone 2010]).

Extending the Mind: Cognitive Extension and the Current Web

The extended mind hypothesis encourages us to think about the way in which much of our cognitive success is grounded in processing loops that factor in the contributions of our extra-neural social and technological environments. In this sense, it is a hypothesis that is highly applicable to discussions about the potential impact of the Web on the human mind. When we apply the concept of the extended mind to the specific socio-technical context of the Web, we end up with the idea of the "Web-extended mind" (Halpin, Clark, and Wheeler 2010; Smart et al. 2010), or the idea that the informational and technological elements of the Web can, at least on occasion, constitute part of the material supervenience base for (at least some of) a human agent's mental states and processes.

Of course, just because we can conceive of something like a Web-extended mind this does not make the realization of Web-based forms of cognitive extension a practical possibility. We need to consider the specific features of the Web and ask whether it constitutes a suitable target for cognitively relevant bio-technological mergers; in particular, whether it meets the kind of criteria encountered in the previous section. At first blush, the answer to this question seems straightforward: given the thought experiment used by Clark and Chalmers (1998)— the one involving Otto, Inga, and the notebook—we can easily imagine a scenario in which the notebook is supplanted with a more technologically sophisticated resource, such as a Web-enabled portable device (see the example scenario discussed in Smart et al. 2010). Such a device would seem to provide much of the same functionality as a notebook, and, in some cases, it might even provide additional functionalities that obviate some of the philosophical criticisms of the extended mind hypothesis.[1] Weiskopf (2008), for example, has criticized the extended mind hypothesis on the grounds that a conventional notebook fails to deliver the kind of informational updating capabilities that we normally expect in the case of biologically based belief states. However, while Weiskopf may be right about the shortcomings of a paper notebook, it is not clear that such criticisms have the same leverage when we think about more sophisticated technological resources (see Smart 2010).

Another advantage of Web-based access is, of course, that we are put in touch with a vast repository of information and knowledge—one much larger than anything we could hope to accumulate in a conventional notebook. The implications of this informational access from an extended mind perspective are potentially profound. For example, if we are enabled to have more-or-less immediate, reliable, and easy access to bodies of information on the Web, and if such information is indeed poised to count as part of our own body of knowledge and beliefs about the world—in the same way, perhaps, as the content of our biologically based (semantic) memories—then we may be only a few technological steps away from an era in which the limits of our personal knowledge are defined by the extent of the Web's reach!

The problem with this upbeat vision, however, is that at present it is unclear to what extent the Web (as currently constituted) possesses the kind of features that would support the emergence of Web-extended minds. The primary problem is that most forms of cognitive extension depend on a particular form of information flow and influence in which there is a close temporal coupling between the various elements that compose the extended system. In the case of the extended mind, for example, it seems reasonable to insist that there should be some functional similarity in the influence exerted by both bio-external and bio-internal

[1] One concern that is sometimes raised about the proposed substitution of Otto's notebook with a Web-enabled device is that the Web does not support the same kind of content editing capabilities that are seen in the case of a conventional notebook. While this was true of earlier versions of the Web, the advent of Web 2.0 has essentially taken us a step closer to the era of the read-write Web, and it is fair to say that most of the content available on the Web today is generated by users posting and uploading content.

information sources. If external information should exert an influence on our thoughts and actions that is profoundly unlike that seen in the case of (for example) bio-memory, then it seems unlikely that we approach the kind of conditions under which genuine forms of mind-extending bio-technological integration take place. And if we think about the kind of informational contact we have with the conventional Web (the Web of HTML [hypertext markup language] documents), then it seems unlikely that we will ever have a form of contact in which information can be accessed in a way that resembles the contents of our long-term memories (see Smart et al. 2009). One area of concern in this respect is the current reliance on document-centric forms of information representation. Ever since the invention of the Web, the dominant way of representing infor-mation content has been via the use of HTML. Traditionally, information has been delivered in the form of HTML documents, which are accessed by browser technologies and then presented to human users in the form of Web pages. This page, or document-centric, mode of representation has significant implications for how we access information content, and it affects the kinds of influence that Web-based information can exert over our thoughts and actions. If we want to enable the kinds of information flow and influence that support the emergence of extended minds, then we need to ensure that our contact with the Web fulfils the kind of criteria outlined by Clark (i.e., the criteria of accessibility, availability, and trust). It is by no means clear, however, that our current reliance on document-centric modes of information delivery really do enable us to meet these criteria. As an example of the shortcomings of document-centric modes of information representation, think about the problem of accessing factual information from a Web-accessible resource, such as Wikipedia. Even if the delays associated with document retrieval and presentation are resolved, the user is still confronted with the onerous task of surveying the document for relevant information content. In most cases, this requires the user to scroll through the Web page and process large amounts of largely irrelevant content in order to identify the small amount of infor-mation that is actually needed. This is a very inefficient means of infor-mation access. Even if the user tries to isolate specific information items for use on multiple occasions, he or she cannot do this without reliably fixing the physical location of the information (perhaps by copying the required information to a local resource). Ideally, what is required in order for Web-based content to count as part of an extended mind is that the relevant factual content should be available to guide thought and action in the ways we have come to expect our thoughts and actions to be guided by information retrieved from bio-memory. The problem with the conven-tional, document-centric Web—the Web of Documents—is that the relevant pieces of information that are required to guide, scaffold, and constrain our thinking are usually embedded in a mass of other distract-ing information. This makes it difficult to see how current forms of

Web-based content could have the kind of functional poise sufficient to count as part of our personal body of knowledge and beliefs about the World.

This is not to say, of course, that the Web does not have the *potential* to serve as the target of cognitively significant bio-technological mergers. The Web is something of a protean beast when it comes to user interaction and information access capabilities. New forms of information representation, new forms of user interaction technology, and new forms of application development all contribute to an ever-changing landscape against which our notions of the Web's capacity to support cognitive extension are always likely to be somewhat ephemeral. In addition to this, we also need to consider the possibility that cognitive extension may depend on the emergence of specific social conventions and practices that determine how the *social* use of a technology is optimized to support cognition at both the individual and the collective levels. The Web is a relatively recent technology, and it is still undergoing rapid development and change. In the next section, I argue that cognitively potent forms of biotechnological merger do not necessarily come for free. They are often the end product of a complex, co-dependent, process of technological innovation, social change, and even neurological configuration. The relatively recent emergence of the Web means that we should not expect it to immediately fulfil all of our requirements in terms of its capacity for cognitive extension. It sometimes takes time for the true transformative potential of a technology to be fully realized.

Socio-Technical Evolution and the Making of an Extended Mind

In the introduction to his book on cognitive extension entitled *Supersizing the Mind: Embodiment, Action, and Cognitive Extension*, Andy Clark recounts an exchange between the Nobel Prize–winning physicist Richard Feynman and the historian Charles Weiner in which Feynman argues for the importance of his notes in contributing to his thoughts. Feynman seems to be arguing that rather than representing a mere record of his internal cogitation, the process of creating the notes and sketches is part of the cognitive work itself. Clark agrees. He argues that the cycle of information flow and influence established by Feynman and his notepad plays a crucial role in the realization of Feynman's thinking: "I would like to go further and suggest that Feynman was actually thinking on the paper. The loop through pen and paper is part of the physical machinery responsible for the shape and flow of thoughts and ideas that we take, nonetheless, to be distinctively those of Richard Feynman" (2008, xxv).

What Clark is basically suggesting here is that writing is constitutive of thinking: that writing plays an active role in the realization of our thoughts, and that the machinery of cognition extends to include not just the biological brain but also the elements of the bio-external environment that make the creation of symbolic artefacts (written words) possible.

The cognitive virtues of writing are a common focus of attention for those interested in technology-mediated cognitive enhancement (e.g., Dix 2008). Clark (1997), for example, suggests that the process of writing an academic paper is one that involves the integration of a variety of props, aids, and artefacts into complex nexuses of environmentally extended information processing. "Extended intellectual arguments and theses," Clark argues, "are almost always the products of brains acting in concert with multiple external resources" (1997, 207).

Writing therefore emerges as a compelling example of technology-mediated cognitive empowerment. Such empowerment, however, does not come for free. The fact is that almost all forms of technologically based cognitive extension require a prolonged period of technological, sociological, and even neurological adaptation. In the case of writing, the features that make writing technologies so apt to participate in extended cognitive systems have come at the end of a long process of technological innovation and social change. The widespread availability of pen and paper, for example, did not happen overnight. It took many years before these resources were available in sufficient quantity for them to become a standard part of our everyday lives—part of the persisting backdrop against which our current set of cognitive capabilities emerged. And although technological innovation and adoption are clearly a major part of the story, they are not the only things that need to be considered. We are not born skilled writers, capable of wielding a pen or working a keyboard. Rather, our writing abilities emerge over the course of many years of instruction and training, often undertaken as part of a formal education. Pen and paper may be simple technologies, but their proper use and exploitation come only at the end of a rather prolonged period of socially scaffolded neurological adaptation and configuration. Bio-technological bonding does not necessarily come for free: we often need to teach our brains how to press maximal cognitive benefit from the technologies we use.

The history of writing also teaches us about the importance of social practices and conventions in enabling a technology to realize its full potential. In using writing technology, we follow a socially accepted set of principles and guidelines governing the proper use of those technologies, and, over time, those principles have evolved to enable the technology to meet its designed purpose. The importance of these social conventions is apparent when we look at early forms of writing. These were not necessarily conducive to cognitive enhancement in the way that later forms were. In fact, early forms of writing were heavily influenced by the traditions and practices of the preceding era, within which information was communicated by purely oral means. Writing initially assumed the form of a continuous stream of text (known as *scriptura continua*), which was devoid of any of the conventional orthographic features (such as word spacing and punctuation) that we now accept as standard features of a written text. The reason for this particular form of writing seems to be based on the fact that writing was initially seen as a means to record and re-present orally communicated information. It simply never occurred to the early writers that the new system of recording thoughts and ideas could be used independently of the spoken word: "It's hard for us to imagine today, but no spaces separated the words in early writing. In the books inked by scribes, words ran together without any break across every line on every page, in what's now referred to as *scriptura continua*. The lack of word separation reflected language's origins in speech. When we talk, we don't insert pauses between each word—long stretches of syllables flow unbroken from our lips. It would never have crossed the minds of the first writers to put blank spaces between words. They were simply transcribing speech, writing what their ears told them to write" (Carr 2010, 61). Scriptura continua, therefore, seems to reflect an intermediate stage in the transition from oral to written culture. It was a form of writing that was heavily influenced by previous forms of information communication, and, like many innovations, it took time for it to become optimally suited to its target audience.

This history of writing serves as a fitting backdrop to the present discussion on Web-extended minds because it helps us understand why current incarnations of the Web may lack the kinds of features that support cognitive extension. We saw that early forms of writing, like the use of scriptura continua, were heavily influenced by the preceding oral tradition. It took some time before the social practices and conventions associated with writing were optimized to take account of its new cognitive role. Like early forms of writing, I suggest that the current form of the Web (the Web of HTML documents) is heavily influenced and informed by the practices and conventions of the era that preceded it: the era of the written and printed word. When information is published on the Web, it is typically done so in the form of "pages" that communicate information in much the same form as we would expect to see in a printed document. This mode of information presentation is, like early forms of writing, not necessarily best suited to the emergence of extended cognitive systems. Perhaps the current version of the Web is, like scriptura continua, an initial form of a technology that is attempting to free itself of the metaphors of a previous era and evolve into something that is far more suited to the realization of its true cognitive potential. What the history of writing teaches us is that we should not mistake the early forms of a technology as constituting the final word in terms of that technology's ability to transform our cognitive and epistemic potential. The making of an extended mind is not something that is necessarily rapid or straightforward: it sometimes takes a long time for a technology to be available in the right form, and we sometimes need to engage in extensive training before we can derive maximum cognitive benefit from the use of a technology. In addition to this, the use of a technology is often guided by social conventions, and these need to be carefully aligned with both the form of the technology and our ability to use it (or our ability to learn how to use it).

Extending the Mind: Cognitive Extension and the Future Web

The transition to a Web that is capable of supporting the emergence of Web-extended minds arguably requires a number of innovations, and together these innovations provide us with an alternative vision of the Web and the kind of interactive opportunities it affords. In this section, I describe two areas of technology development that are already beginning to change our relationship to Web-based information content. The first of these areas concerns the use of linked data formats to change the way information content is represented on the Web (Bizer, Heath, and Berners-Lee 2009). Such formats improve both the accessibility (in terms of retrieval of specific pieces of relevant information) and the versatility (in terms of flexible modes of presentation) of information content. The second area of technology development concerns the use of new kinds of display device and augmented reality capabilities. These change the nature of our relationship to information on the Web by making that information more accessible and more suitably poised to influence our everyday thoughts and actions.

The Missing Link: Towards the Web of Data

We have seen that the conventional Web—the Web of HTML documents, or Web of Documents—presents a number of problems for Web-based forms of cognitive extension (for example, the difficulties of accessing relevant information that is embedded in a mass of other, semantically irrelevant, content). Fortunately, an alternative approach to the representation of online content is emerging alongside the conventional Web of Documents. This is the Web of Data (Bizer, Heath, and Berners-Lee 2009), which is based on the idea that the Web should serve as a globally distributed database in which data linkages are established by dereferenceable URIs (uniform resource identifiers). The transition from document-centric to data-centric modes of information representation is, I think, an important step in the technological evolution of the Web, particularly when it comes to the notion of the Web-extended mind.[2] What is important for Web-based forms of cognitive extension is flexible modes of data

---

[2] Paradoxically, although the aim of the linked data initiative is to provide content that is primarily suited for machine-based processing, while the aim of the conventional Web (the Web of Documents) is to provide content suited for machine-based processing, it is the Web of Data that, I suggest, provides the better opportunity for human-centred cognitive transformation.

integration, aggregation, filtering, and presentation, in conjunction with an ability to gear information retrieval operations to suit the task-specific needs and requirements of particular problem-solving contexts. The Web of Data supports these capabilities in a number of ways. In particular, it countenances representational formats that are:

1. largely independent of specific presentational formats or usage contexts (this supports flexibility in the way information content is presented; it also enables data to be used in different ways in different application contexts),
2. centred on the representation of limited sets of data or data items (this supports the rapid retrieval and presentation of specific pieces of information), and
3. semantically enriched (this supports the retrieval of relevant information).

This mix of features brings us a step closer to establishing the kind of conditions under which the emergence of Web-extended minds is a realistic possibility. The Web of Data is not necessarily the final step in this process, but it does provide an important step, I think, towards more potent (and empowering) forms of cognitive engagement with the Web.

The Real World Web

The gradual transition from document-centric to data-centric modes of information representation is one of the ways in which the Web is evolving to provide new opportunities for cognitive augmentation and enhancement. Without the correlative development of suitable interaction mechanisms, however, the potential for the new representational formats to genuinely transform our cognitive and epistemic potential is still somewhat limited. Although there are new types of browsers that enable human users to browse the Web of Data, it is unlikely, I think, that these browsers will introduce any radically different forms of informational contact, at least relative to the kind of contact already afforded by conventional Web browsers. Rather than focus on the development of browser interfaces that simply take existing functionality and adapt it for the Web of Data, I suggest that we need to think about radically new forms of information display and user interaction. We need to move beyond the browser interface, which for too long has dominated our notions of informational contact with Web-accessible information resources. Instead, I suggest that we need to entertain a new vision of the Web: one that makes bio-external information resources suitably poised to participate in the emergence of Web-extended minds. Let us refer to this new vision of the Web as the "Real World Web."

As is suggested by its name, real-world environments are at the heart of the concept of the Real World Web. The basic idea is that Web-based information should, wherever possible, be embedded in the real world and easily accessed as part of our everyday interaction with that world. Information about everyday objects should be associated with those objects, information about locations should be accessible in those locations, and information about people should be "attached" to those people. In all cases, the information should be immediately accessible and easily processed. It must be able to guide thought and action in the way in which our everyday cognition is supported by the information retrieval operations of our own biological brain. What this means, in effect, is that information should be immediately accessible to our perceptual systems. It should require little or no effort to make the information available for perception, and, in most cases, the information should be delivered automatically, with little or no conscious effort required to make that information available.

This vision is one which modifies our traditional modes of interaction with the Web in a number of ways. Instead of the retrieval of relevant information being entirely the responsibility of the human agent, the notion of the Real World Web advocates a more intelligent and proactive Web: a Web that is capable of anticipating users' information requirements and making that information available in ways that support cognitive activity. It is also a vision that places the Web at the heart of our everyday embedded interactions with the world. Rather than information access requiring perceptual detachment and disengagement from our immediate surroundings (something that is required even with the most portable of mobile devices), the notion of the Real World Web seeks to make Web-based information access a standard feature of our everyday sensorimotor engagements with the world—it seeks to make the Web part of the perceptual backdrop against which our everyday thoughts and actions take shape. Finally, the notion of the Real World Web emphasizes a shift away from traditional browser-based modes of Web access, featuring the use of screen-based displays, keyboard-based interaction mechanisms, and document-centred representational schemes. In place of conventional screen-based modes of information access, the Real World Web emphasizes the importance of more perceptually direct forms of information access (e.g., the use of augmented reality devices to overlay Web-based information onto real-world objects and scenes); in place of conventional user interaction devices, such as mice and keyboards, the Real World Web advocates the use of alternative interaction mechanisms (more on which below); and in place of conventional document-centric modes of information representation, the Real World Web countenances a transition to more data-centric modes of information representation (see above). The main implication of this shift away from conventional browser-based modes of Web access is that we are enabled to see the Web in a new light: as something more than a passive set of information resources that need to be coaxed into useful cognitive service by our deliberate search and retrieval efforts. Our traditional modes of access with the Web encourage us to see the Web as something that is:

1. *Passive:* we need to engage with the Web in a deliberate manner in order to retrieve relevant information. Information needs to be discovered, retrieved, filtered and interpreted; seldom does the Web support us in a proactive manner—providing the right information just when we need it.

2. *Distinct from our everyday interaction with the world:* the Web may support our everyday decision-making and problem-solving behaviours, but, in general, this support comes at the cost of us having to divert our attention away from the problem at hand. Instead of information being immediately available to support our thoughts and actions, we are often required to suspend what we are doing in order to look things up on the Web.

3. *Impersonal:* the information content of the Web is, in general, not geared to specifically suit our idiosyncratic problem-solving needs and concerns; often we have to access information from several sources and adapt it for our own ends.

In place of this vision, the Real World Web gives us a vision of the Web as something that is proactive, personal, and perceptually immediate. Once we are afforded immediate perceptual access to Web-based information, and once such information becomes available at just the right time to support our goals, interests, and concerns, then such information becomes, I suggest, far more capable of fulfilling the kind of conditions that merit the emergence of Web-extended minds.

The Real World Web, it should be clear, relies on a rich range of sophisticated technical capabilities, most of which are, as yet, either unavailable or not in widespread use. This might be perceived as grounds for pessimism about the tenability of the Real World Web vision. For the most part, however, the kind of technological innovations required to make the Real World Web a reality are not beyond the reach of our current engineering capabilities, and, in some cases, early forms of the technologies are already starting to appear. One of the most interesting and relevant areas of recent technological innovation concerns the development of a range of highly portable augmented reality or mixed reality solutions. These are available in a variety of forms, from handheld mobile devices that overlay information onto a real-world scene via a camera and screen display, through to wearable-computing solutions, such as head-mounted displays and retinal projection systems. One such system, which is being developed by researchers at the University of Washington, uses a set of micro-fabrication techniques to incorporate display micro-devices into a contact lens (Lingley et al. 2011). The contact lens is worn like any other contact lens, and it provides a see-through display that is both remotely powered and controlled via a wireless link. Another device that has attracted recent media attention is the Internet-connected head-mounted display system envisioned by Google's Project Glass initiative (Bilton 2012). The ultimate promise of such devices is that they enable network-accessible information to be superimposed on real-world objects and scenes, enhancing the kind of informational contact we have with our online world, and significantly enriching the range of perceptual cues, prompts, and affordances that guide our everyday thoughts and actions.

Conclusion

Throughout history, technologies have exerted a significant influence on humanity. Some technologies have contributed to profound forms of socio-economic change, fundamentally transforming the nature of our social activities and modifying the structure of our social engagements. Others have contributed to a profound change in our cognitive profiles, fundamentally transforming the way we think about the world and modifying the basic character of our cognition. Like many of the technologies that preceded it, the Web is a technology that is potentially poised to further transform both society and ourselves. At present, much of the research related to the Web has focused on its social and societal impact: the effect the Web is having on social, political, and cultural processes. The cognitive impacts of the Web, however, are also an important focus of scientific and philosophical attention. The present chapter has explored the notion of the Web-extended mind—the idea that the informational and technological elements of the Web may support environmentally extended cognition in the manner suggested by the extended mind hypothesis (Clark and Chalmers 1998). Following a consideration of the Web's features in relation to a number of criteria outlined to discriminate genuine cases of cognitive extension from more ersatz varieties, it was concluded that the Web is not particularly well suited to the formation of extended cognitive systems.

We should not, however, be unduly surprised or downcast by this conclusion. When thinking about the cognitive impact of the Web it is always important to be clear that the Web is a relatively recent technology and is still undergoing rapid development and change. This means that our conception of what the Web is is a highly dynamic one. Current modes of interaction with the Web do not necessarily limit (or even limn) the space of interactive opportunities that could be created by future forms of technological innovation, and we thus need to be wary of blanket statements about the Web's ability to help or hinder cognitive processing. In addition to this, we saw in the case of writing that some technologies take time to become ideally suited to supporting cognitive extension. Perhaps something similar is true of the Web. As was the case with writing, our ability to gain maximal cognitive benefit from the Web will depend on a progressive optimization of the technologies we use and the social conventions that dictate how we use them. The current Web is surely an important point in our cultural and intellectual history, but it need not be the final word in terms of our cognitive and epistemic transformation. Perhaps, with time, we will come to see the current Web as laying the groundwork for more potent and profound forms of cognitive enhancement—an important, albeit temporary, milestone en route to our Web-extended cognitive destiny.

References

Adams, Frederick, and Kenneth Aizawa. 2008. *The Bounds of Cognition.* Oxford: Blackwell.

Bilton, Nick. 2012. "A Rose-Colored View May Come Standard." *New York Times* (5 April), available at http://www.nytimes.com/2012/04/05/technology/google-offers-look-at-internet-connected-glasses.html

Bizer, Christian, Tom Heath, and Tim Berners-Lee. 2009. "Linked Data: The Story so Far." *International Journal on Semantic Web and Information Systems* 5, no. 3:1–22.

Carr, Nicholas. 2010. *The Shallows: How the Internet Is Changing the Way We Think, Read and Remember.* London: Atlantic Books.

Clark, Andy. 1997. *Being There: Putting Brain, Body and World Together Again.* Cambridge, Mass.: MIT Press.

———. 1999. "An Embodied Cognitive Science." *Trends in Cognitive Science* 3, no. 9:345–51.

———. 2008. *Supersizing the Mind: Embodiment, Action, and Cognitive Extension.* New York: Oxford University Press.

———. 2010. "Memento's Revenge: The Extended Mind, Extended." In *The Extended Mind,* edited by Richard Menary, 43–66. Cambridge, Mass.: MIT Press.

Clark, Andy, and David Chalmers. 1998. "The Extended Mind." *Analysis* 58, no. 1:7–19.

Dix, Alan. 2008. "Externalisation: How Writing Changes Thinking." *Interfaces* 76:18–19.

Halpin, Harry, Andy Clark, and Michael Wheeler. 2010. "Towards a Philosophy of the Web: Representation, Enaction, Collective Intelligence." In *Proceedings of the 2nd Web Science Conference,* 26–27 April 2010, Raleigh, North Carolina, USA, available at http://journal.webscience.org/324/

Haugeland, John. 1998. "Mind Embodied and Embedded." In *Having Thought: Essays in the Metaphysics of Mind,* edited by John Haugeland, 207–37. Cambridge, Mass.: Harvard University Press.

Hurley, Susan. 1998. *Consciousness in Action.* Cambridge, Mass.: Harvard University Press.

Hutchins, Edwin. 1995. *Cognition in the Wild*. Cambridge, Mass.: MIT Press.

Lingley, A. R., M. Ali, Y. Liao, R. Mirjalili, M. Klonner, M. Sopanen, S. Suihkonen, T. Shen, B. P. Otis, H. Lipsanen, and B. A. Parviz. 2011. "A Single-Pixel Wireless Contact Lens Display." *Journal of Micromechanics and Microengineering* 21, no. 12:125014–21.

Menary, Richard, ed. 2010. *The Extended Mind*. Cambridge, Mass.: MIT Press.

Pfeifer, Rolf, and Josh Bongard. 2007. *How the Body Shapes the Way We Think: A New View of Intelligence*. Cambridge, Mass.: MIT Press.

Robbins, Phillip, and Murat Ayded. 2009. *The Cambridge Handbook of Situated Cognition*. New York: Cambridge University Press.

Rowlands, Mark. 1999. *The Body in Mind*. New York: Cambridge University Press.

———. 2006. *Body Language: Representation in Action*. Cambridge, Mass.: MIT Press.

———. 2010. *The New Science of the Mind: From Extended Mind to Embodied Phenomenology*. Cambridge, Mass.: MIT Press.

Rupert, Robert. 2009. *Cognitive Systems and the Extended Mind*. New York: Oxford University Press.

Smart, Paul R. 2010. "Extended Memory, the Extended Mind, and the Nature of Technology-Mediated Memory Enhancement." In *1st ITA Workshop on Network-Enabled Cognition: The Contribution of Social and Technological Networks to Human Cognition*, 21 September 2009, Adelphi, Maryland, USA, available at http://eprints.soton.ac.uk/267742/

Smart, Paul R., Paula C. Engelbrecht, Dave Braines, Mike Strub, and Cheryl Giammanco. 2010. "The Network-Extended Mind." In *Network Science for Military Coalition Operations: Information Extraction and Interaction*, edited by Dinesh Verma, 191–236. Hershey, Penn.: IGI Global.

Smart, Paul R., Paula C. Engelbrecht, Dave Braines, Mike Strub, and Jim Hendler. 2009. "Cognitive Extension and the Web." In *Proceedings of the 1st Web Science Conference*, 18–20 March 2009, Athens, Greece, available at http://eprints.soton.ac.uk/267155/

Theiner, Georg, Colin Allen, and Robert L. Goldstone. 2010. "Recognizing Group Cognition." *Cognitive Systems Research* 11:378–95.

Tollefsen, Deborah P. 2006. "From Extended Mind to Collective Mind." *Cognitive Systems Research* 7, nos. 2–3:140–50.

Weiskopf, Daniel A. 2008. "Patrolling the Mind's Boundaries." *Erkenntnis* 68, no. 2:265–76.

Wilson, Robert A. 2000. "The Mind Beyond Itself." In *Misrepresentations: A Multidisciplinary Perspective*, edited by Dan Sperber, 31–52. New York: Oxford University Press.

———. 2004. *Boundaries of the Mind: The Individual in the Fragile Sciences: Cognition*. New York: Cambridge University Press.

Wilson, Robert A., and Andy Clark. 2009. "Situated Cognition: Letting Nature Take Its Course." In *Cambridge Handbook of Situated Cognition*, edited by Murat Ayded and Phillip Robbins, 55–77. Cambridge: Cambridge University Press.

CHAPTER 9

GIVEN THE WEB, WHAT IS INTELLIGENCE, REALLY?

SELMER BRINGSJORD AND
NAVEEN SUNDAR GOVINDARAJULU

1. The Web Extended and Immediate: True Intelligence?

Let's assume that vast declarative information covering nearly all human collective knowledge, courtesy of the Semantic Web a decade hence,[1] enables a flawless version of what is known as **QA** technology.[2] Add a second assumption: namely, that the QA cycle is mediated by direct brain-Web interfacing. Under this foreseeable-in-the-near-future assumption, if Smith is verbally asked a question q, he can internally and mentally ask it of the Semantic Web, receive back an answer a immediately to his neocortex, and convey a as required (e.g., by vocalizing the answer to the interlocutor before him). If Smith could do this today, surreptitiously, he would certainly cause most questioners to believe that he is rather intelligent. He would, for example, be able to: say which of Shakespeare's plays contain any given snippet of the Bard's immortal narratives (e.g., "What did devout Harry's former friend steal, and pay for with his life?"), answer any question about any settled part of science (e.g., "What was Frege's quirky notation for what is essentially modern first-order logic?"), produce the notes in sequence for Bach's German Organ Mass (and throw in a word-for-word verbalization of Luther's Catechism), and so on. If no one knew about Smith's hidden, wireless, brain-to-Web link, again, he certainly would be regarded intelligent—probably even positively brilliant, at least by many.

2. Cartesian Skepticism

By many, but not by all. What would a truly discriminating judge say? We're afraid that on accounts of *real* intelligence of the sort that Descartes had in mind, your secret Web link would be insufficient. Why? Because

---

[1] In the present chapter we use the terms "Semantic Web" and "Web" interchangeably.
[2] Undeniably, the best QA technology in the world is currently the Watson system, created by IBM to compete against humans in the long-running American television game *Jeopardy!* which is essentially itself a QA game. For a description of the system, see Ferrucci et al. 2010. Given Watson's prowess, the future we envision could very well soon arrive.

Philosophical Engineering: Toward a Philosophy of the Web, First Edition. Edited by Harry Halpin and Alexandre Monnin. Chapters © 2014 The Authors except for Chapters 1, 2, 3, 12, and 13 (all © 2014 John Wiley & Sons, Ltd.). Book compilation © 2014 Blackwell Publishing Ltd and Metaphilosophy LLC. Published 2014 by Blackwell Publishing Ltd.

Descartes would know that a mere mechanical machine could in principle do just what you qua QA master are doing. Thus, by considering whether the Web, given current trends, will fundamentally alter the very concept of human intelligence, we find ourselves carried back to the long-standing debate about whether human (or human-*level*) intelligence can be captured in mechanical form. Descartes answered this question in the negative. Long before Turing, he claimed that only the human has *domain-independent* (and conversational) intelligence, and that therefore certain tests would be exceedingly difficult, if not outright impossible, for machines to pass. He specifically suggested two tests to use in order to separate mere machines from human persons. The first of these directly anticipates the so-called Turing Test. The second test is the one that connects directly to domain-independent intelligence. We read:

> If there were machines which bore a resemblance to our body and imitated our actions as far as it was morally possible to do so, we should always have two very certain tests by which to recognise that, for all that, they were not real men. The first is, that they could never use speech or other signs as we do when placing our thoughts on record for the benefit of others. For we can easily understand a machine's being constituted so that it can utter words, and even emit some responses to action on it of a corporeal kind, which brings about a change in its organs; for instance, if it is touched in a particular part it may ask what we wish to say to it; if in another part it may exclaim that it is being hurt, and so on. But it never happens that it arranges its speech in various ways, in order to reply appropriately to everything that may be said in its presence, as even the lowest type of man can do. And the second difference is, that although machines can perform certain things as well as or perhaps better than any of us can do, they infallibly fall short in others, by which means we may discover that they did not act from knowledge, but only for the disposition of their organs. For while reason is a universal instrument which can serve for all contingencies, these organs have need of some special adaptation for every particular action. From this it follows that it is morally impossible that there should be sufficient diversity in any machine to allow it to act in all the events of life in the same way as our reason causes us to act. (Descartes 1911, 116)

Figure 9.1 offers an illustration of what Descartes expresses in the quote immediately above.

A test of domain-independent intelligence requires success on topics with which the agent has had no prior experience. In our thought experiment, by definition, you look smart specifically because you are really you *plus* the Web, and the Web gives you access to mountains of *preestablished* information—but not to "knowledge" in Descartes's sense of the word here. What he means by knowledge in this context is a capacity for general-purpose, problem-solving knowledge.

To follow Descartes in testing for real intelligence we must present you with a problem that you have never seen before, and wait to see whether

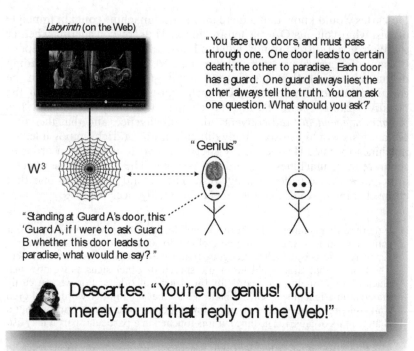

FIGURE 9.1. Using the Web to find declarative knowledge

you can provide a solution by means that you invent on the spot.[3] We hold that while the Web and associated cognitive technologies promise to make possible the fake brilliance of Smith, the Web will not provide the "universal instrument" that resides within us. Hence the Web won't bring us any closer to the flexible, general intelligence that Descartes correctly claimed would continue to separate minds from mere machines, as figure 9.2 illustrates.

3. The Missing Science of Human+Web Intelligence

Even those who disagree with us must admit two things: one, our position has *some* force; and two, this very fact, combined at the same time with reluctance to wholeheartedly affirm our position, implies that the matters before us are fundamentally unsettled; or to put that point another way: there simply isn't currently available a rigorous science of human-

---

[3] A series of simple yet remarkably subtle tests of this form, for children and adolescents, are provided by Piaget (Inhelder and Piaget 1958).

FIGURE 9.2. Using the Web to find imperative knowledge

intelligence-augmented-by-the-Web (H+W). Addressing the absence of such a science is nontrivial, as we'll soon proceed to show. To encapsulate here: Today, all knowledge represented formally and hence amenable to automated reasoning on the Semantic Web is extensional; but human reasoning is extensional *and* intensional.[4] The problem stems from a lack of formal computational systems that can handle intensional reasoning as robustly as they can handle extensional reasoning. The formal and computational sciences are in the very early stages of understanding such systems, and here we briefly explore some of the initial moves in a research program aimed at producing a formal science of H+W.

4. A Desideratum for a Science of H+W: Modeling Knowledge

Most human reasoning is highly intensional, in that it involves belief, desire, knowledge, action, intention, perception, communication, and so

---

[4] The terms "extensional" and "intensional" derive from logic. Put barbarically, extensional logics are concerned merely with the references of terms, while intensional logics are concerned with both the sense and the reference of a term. In concrete practice, the student of mathematical logic learns first the simplest extensional logics (propositional calculus, first-order logic, for example), and will only be confronted with the need to invoke intensional operators when taught philosophical logic. A nice overview of the latter class of logical systems is provided in Goble 2001.

on. This is in opposition to standard reasoning recorded, for instance, in mathematics or the natural sciences; such reasoning is extensional. Consider the following two statements:

P_1: My infant niece does not know that the area of a circle with radius r is $\pi * r^2$

P_2: The area of a circle with radius r is $\pi * r^2$

The first statement is intensional, the second is extensional. Why do we need to worry about statements of the first type in order to erect a rigorous science of H+W?

Consider this scenario: Jack is married to Jill, and in preparation for their anniversary, he **communicates** his **intention** of planning a surprise dinner for her—through the brain-computer interface chip implanted in his brain—to his personal assistant Simon in the **hope** that Simon will take care of his **planning**. Jack also **believes** that, even though he has not **communicated** to Simon that Jill should not know of Jack's **intentions**, Simon **knows** that this should not be made **known** to Jill until the evening before the dinner.

Even this perfectly ordinary scenario demonstrates an *explosion of intensionality* that no current theory is capable of handling—despite impressive progress in AI, and in Web-powered QA and Web-powered personal assistants. Also, note that this task includes what could be general-purpose planning, still an unsolved problem. Further complicating the matter is the presence of indexical terms: *this, he, his, their*, and so on.

All attempts at formalizing reasoning on the Web have so far dealt only with shallow and nongeneral reasoning over extensional data, reasoning woefully short of being able to handle the above scenario.[5] What is necessary to rectify this? The answer is clear: a system able to deal with intensionality, ranging across all intensional operators that appear in scenarios like the one above involving Jack and Jill. While classical mathematical logic has been applied successfully to extensional domains, the situation for the intensional case is not so bright, as we show below. This state of affairs imperils any serious attempt at understanding, scientifically, Web-enhanced human intelligence.

We present three possible ways of tackling just one kind of intensionality, one that takes us back to the starting parable of Smith: *declarative knowledge possessed by an agent*; and we show how these ways all lack robustness. We seek to lend credence to an approach based on intensional modal operators but are happy to see a future unfold in which intensional operators in English are attacked in any reasonable manner.

---

[5] In fact, though we don't discuss it, the Semantic Web is currently quite firmly tied to not only extensional logics but also but low-expressivity ones, e.g., description logics. Such logics are covered in Baader, Calvanese, and McGuinness 2007.

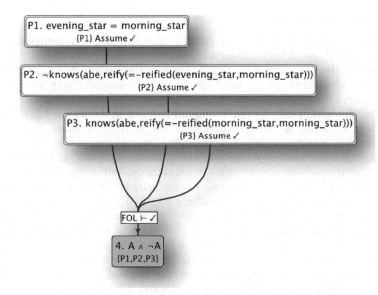

FIGURE 9.3. An attempt at modeling knowledge in a first-order language

4.1. A First Attempt

An initial attempt at modeling declarative knowledge might proceed to do so by incorporating a knowledge *predicate* directly into first-order logic. One way to go about this is by mirroring all predicate symbols with another set of new function symbols. For instance, a predicate symbol **P** gets mirrored by the function symbol f_P; this process is called *reification* or, more aptly given how it is often implemented (e.g., see Russell and Norvig 2002), *stringification*.

Unfortunately, this approach is doomed from the start. Consider the small and commonly known illustration in figure 9.3, implemented in the Slate proof-engineering environment (Bringsjord et al. 2008). Our hypothetical person, Abe, is quite ignorant of modern astronomy and doesn't know that the morning star is the very same object as the evening star. They are indeed the same object: the planet Venus. This is a fact that was unknown to many of our ancestors; they saw the planet at dawn and dusk and believed they were different objects. Any honest modeling formalism must respect this state of affairs, but alas, we find that as a soon as we enter the relevant propositions into Slate we encounter a contradiction. We still have a problem if it was unknown to us that Abe does not know that the morning star is in fact the evening star: our inference becomes unsound.

4.2. A Second Attempt

One way to escape from the contradiction and unsoundness pointed to by the proof in figure 9.3 is to devise a knowledge predicate that applies to the Gödel numbers of formulae, which blocks certain unwanted inferences, namely, substitution of co-denoting terms (e.g., morning_star with evening_star).

This discussion follows the discussion laid out in Anderson 1983. The Knower Paradox is an epistemic paradox that stems from assuming certain obvious premises about knowability. A semi-automated version of the paradox is shown in figure 9.4. The paradox is arrived at as follows: Consider the well-known system Q of Robinson Arithmetic (see Boolos, Burgess, and Jeffrey 2003) along with a predicate $\mathbf{K}$ the extension of which is supposed to be the Gödel numbers of all the sentences of Q that are known to some knower. We are also given another predicate, $\mathbf{I}$, the extension of which is the set of the ordered pairs of Gödel numbers ($\lceil \phi \rceil$, $\lceil \psi \rceil$) such that ψ is derivable assuming ϕ in Q. (We use corner brackets to denote the Gödel number of a formula, and the idiomatic notation with a bar over the top of a formula to denote the Gödel numeral for the formula.) Now, one could reasonably suppose the following axioms for the new predicates:

$$Ax_1 \ \mathbf{K}(\overline{\phi}) \to \phi$$

$$Ax_2 \ \mathbf{K}\left(\overline{\mathbf{K}(\overline{\phi}) \to \phi}\right)$$

$$Ax_3 \ \mathbf{I}(\overline{\phi}, \overline{\psi}) \to \left(\mathbf{K}(\overline{\phi}) \to \mathbf{K}(\overline{\psi})\right)$$

Unfortunately, contradiction still rears its head, as shown in the semi-automated Slate proof in figure 9.4. Note that in this proof, Gödel numerals are represented with a $\triangleright(\phi)$ instead of $\overline{\phi}$.

4.3. A Third Attempt

There are many ways by which one could avoid this paradox, but it seems that all of them are less than satisfactory. One could argue that iterated knowledge predicates don't always have truth values, or that knowledge need not be monotonic (i.e., that old knowledge can be destroyed when new knowledge is acquired). These two suggestions, while bearing the virtue of simplicity, don't accord with our pre-analytic notions of knowledge. Another, more attractive route, suggested by Anderson, is to dissect the predicates $\mathbf{K}$ and $\mathbf{I}$ into a hierarchy of predicates: $\mathbf{K_0}$, $\mathbf{K_1}$, ... , for knowledge; and $\mathbf{I_0}$, $\mathbf{I_1}$, ... , for inference. We obtain the language dubbed L_ω when we augment the first-order language L of arithmetic with these hierarchies of $\mathbf{K}$ and $\mathbf{I}$ predicates. The levels of predicates are supposed to

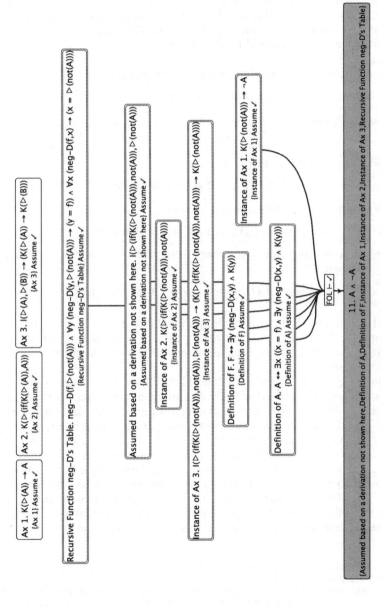

FIGURE 9.4. The Knower Paradox semi-automated

mirror an agent iteratively reflecting upon its knowledge. In this approach, the interpretation V assigns an extension to all the predicates and is also decomposed into a series of interpretations, V_0, V_1, V_2, such that V_i assigns extensions only to $\mathbf{K}_i$ and $\mathbf{I}_i$, and $V = \cup_i V_i$. V_0 extends some interpretation V_p for L to furnish extensions for $\mathbf{K}_0$ and $\mathbf{I}_0$. The following metalogical conditions are then imposed by Anderson:[6]

(i) Knowledge is preserved when reflecting deeper:

$$V_i(\mathbf{K}_i) \subseteq V_{i+1}(\mathbf{K}_{i+1})$$

(ii) Inferences are preserved:

$$V_i(\mathbf{I}_i) \subseteq V_{i+1}(\mathbf{I}_{i+1})$$

(iii) Only true things can be known:

$$\lceil \phi \rceil \in V_i(\mathbf{K}_i) \Rightarrow V_j(\phi) = T \text{ for a } j \geq i$$

(iv) Only correct inferences can be made:

$$(\lceil \phi \rceil, \lceil \psi \rceil) \in V_i(\mathbf{I}_i) \Rightarrow V_j(\phi \rightarrow \psi) = T \text{ for a } j \geq i$$

(v) If something can be inferred from what is known then that is known:

$$(\lceil \phi \rceil, \lceil \psi \rceil) \in V_i(\mathbf{I}_i) \text{ and } \lceil \phi \rceil \in V_i(\mathbf{K}_i) \Rightarrow \lceil \psi \rceil \in V_i(\mathbf{K}_i)$$

Interpretations V that can be decomposed into a hierarchy of interpretations, V_0, V_1, V_2, ..., and satisfying the above conditions, are called *coherent*. A sentence of L_ω is called κ-*valid* iff it is true in every coherent interpretation.

Now, we ask: Is κ-validity semidecidable? Note that κ-validity is just first-order validity strengthened by conditions of coherence on the interpretations. First-order validity is of course semidecidable: If a sentence is valid in FOL, then there is a partial-recursive function p such that $p(x) = 1$ iff x is the Gödel number of a valid first-order sentence (e.g., see Boolos and Jeffrey 1989). The function p can be *represented* in Q;[7] this is so because Q is strong enough to represent all the partial-recursive functions. Let us add a new predicate $\mathbf{P}$ to Q to represent first-order validity; that is, $Q \vdash P(\bar{\phi})$ iff $\{\} \vdash \phi$. Since we are dealing with a provability predicate, we posit the following axioms, assuming that we are concerned with an ideal mathematician who knows all the theorems of Q at the first level of reflection $\mathbf{K}_0$.

---

[6] Note $V(\varphi) = T$ is shorthand for $V \models \varphi$ if φ is a sentence, and $V(P)$ denotes the extension of P if P is a predicate symbol.

[7] Briefly, a function $g: \mathbb{N} \mapsto \mathbb{N}$ is representable in a theory T in the language of arithmetic $L = \{0; =, <; S, +, .\}$ iff there is a formula $G(x,y)$ such that whenever $g(m) = n$ we have $T \vdash \forall x \, G(m,x) \leftrightarrow x = n$. Consult Boolos, Burgess, and Jeffrey 2003 for more background.

A_1: If it is known that the validity of something is proven, then that is known:

$$\mathbf{K}_0\left(\overline{\mathbf{P}(\overline{\phi})}\right) \to \mathbf{K}_0(\overline{\phi})$$

A_2: An instance of Ax_3: (Q_{conj} is the conjunction of all the axioms in Q.)

$$\mathbf{I}\left(\overline{Q_{conj}}, \overline{\mathbf{P}(\overline{\phi})}\right) \to \left(\mathbf{K}_0\left(\overline{Q_{conj}}\right) \to \mathbf{K}_0\left(\overline{\mathbf{P}(\overline{\phi})}\right)\right)$$

A_3: Q_{conj} is known:

$$\mathbf{K}_0\left(\overline{Q_{conj}}\right)$$

A_4: All κ-valid sentences ϕ can have their validity proved in Q:

$$\mathbf{I}\left(\overline{Q_{conj}}, \overline{\mathbf{P}(\overline{\phi})}\right)$$

We can see that from $\{A_1, A_2, A_3, A_4\}$ we have for all κ-valid sentences φ that $\mathbf{K}_0(\overline{\phi})$. We also see that $V_0(\mathbf{K}_0(\overline{\phi})) = T$ for all such φ, since the predicate $\mathbf{K}_0$ is assigned its full extension by V_0. We have the following theorem by Anderson:

Theorem (No i-perfect logician). *No coherent interpretation has an i such that for all κ-valid sentences φ,*

$$V_i\left(\mathbf{K}_i(\overline{\phi})\right) = T.$$

Note that for $i = \mathbf{0}$, the theorem contradicts our result. Our more sophisticated attempts at avoiding contradiction seem to have failed.

Since we have assumed an ideal mathematician in this argument, one could counter that the contradiction obtained is not as strong as the ones obtained before, since—so the objection goes—ideal mathematicians have "obviously" not been observed to exist. Unfortunately, this rejoinder hardly surmounts the problem: the mere *possibility* of the existence of such a mathematician engenders a contradiction, and the mere possibility of such a contradiction is a serious problem for a serious science of knowledge in the context of H+W.

5. Some Objections, and Our Responses

We now present and rebut some of the objections that might be raised against us.

5.1. Objection 1

Objection 1 is essentially a claim that human intelligence, conventionally understood, is not, if you will, "universally" intelligent, as it involves critical use of information/capability encoded in evolved genetic material. Here is one way the objection might be expressed:

> You claim "A test of domain-independent intelligence requires success on topics with which the agent has had no prior experience." Well, I assume that you understand that one could argue that this Cartesian requirement makes sense if one requires "prior experience" to only apply to experience in an individual lifespan. However, one could argue that general purpose-learning is done not via scratch from humans but via "prior experience" encoded in selection of genetic material. So humans do not work in areas in which their species have absolutely no prior experience: Imagine a situation where a human is transported into a completely alien world. He or she would likely die, not adapt.

5.1.1. Our Response to Objection 1. We counter by noting that it is exceedingly hard to see how human reasoning in abstract domains such as logic and mathematics could ever have been encoded in genetic material via the experience of one's ancestors, yet humans reason and solve problems successfully in these abstract domains.[8] It's also hard to see why human persons, if thrust into an alien world, wouldn't retain the power to solve problems there that would yield to the logico-mathematical techniques that work on Earth. Regardless, the *onus probandi* is surely on the critic to provide evidence rather than just an imaginative notion regarding ancestry-bound problem solving. Note that the critic here says: "One could argue." Perhaps that is true enough; but where is the argument? We certainly agree that Objection 1 is indeed imaginative; we furthermore agree that in subsequent dialectic on the formal science of H+W the present exchange is worthy of sustained analysis; but these concessions leave the case we present here quite intact.

5.2. Objection 2

The second objection is that the Semantic Web already uses, or at any rate will soon successfully use, higher-order and intensional logics. It can be expressed as follows:

---

[8] It is interesting to note that Objection 1 is similar to the justification given by Cassimatis 2006 for his fascinating cognitive substrate hypothesis, according to which, in nutshell form, all of our grand cognitive powers today as members of *Homo sapiens sapiens* are reducible to a small core of algorithms and data structures. Presumably all trenchant objections to this hypothesis apply to Objection 1. For instance, logicians and mathematicians can grasp the truth of Goodstein's Theorem and understand its proof. (An accessible account of the theorem and its proof is given by Smith 2007.) The only proofs known of Goodstein's theorems use the ordinal numbers and properties thereof, infinitary concepts and structures that seem to be well beyond a cognitive substrate needed for such pedestrian pursuits as those that defined the lives of hunter-gatherers (Bringsjord 2001).

First, the two of you assume that the Semantic Web is limited to very restrictive logics. This is not true. RDF actually has traditionally higher-order features, such as circular classes and names both referring to classes and properties. Second, first-order logic, and intensional logics, could be added to the Semantic Web in "higher layers" as has been detailed in Tim Berners-Lee's roadmap to a "universal logic." Would simply adding an adequate intensional logic à la Zalta to the Semantic Web allow your thought-experimental Smith to achieve true intelligence, or is there a more properly philosophical argument over prior experience and general-purpose reasoning that prevents the "W" in "H+W" from a priori qualifying as intelligent?

5.2.1. Our Response to Objection 2. Though the formal systems used for the Semantic Web have features that resemble those in higher-order logic and intensional logics, these are but shallow surface-level syntactic commonalities entirely reducible to classical first-order logic; one such reduction is carried out by Fikes, McGuinness, and Waldinger 2002. Even if the Semantic Web community deploys *true* higher-order and intensional logics, the question of the automation of such systems poses an even graver problem.

There is a deeper philosophical argument underlying our reasoning. Our stance is that human persons, in reasoning in and over some domains (at least some *abstract* domains, for example the set-theoretic universe posited in ZF), gain direct and unmediated access to the objects in these domains.[9] Machines, on the other hand, seem to be forever restricted to manipulating syntactic or symbolic structures that merely shadow these abstract objects.[10] Given this, we find it exceedingly hard to see how human intelligence can be genuinely augmented via interfaces with the Web. In sum, our Smith, we still maintain, is no smarter courtesy of the link from his brain to the Web than he is without it. We assume that even obdurate readers will at least concede that in light of the reasoning we have given, the rigorous science of H+W is in its infancy, and, accordingly, much work remains to be done.

6. Summing Up

Because of thought experiments like the one given above involving Smith, the prescientific state of an account of the intelligence of H+W is exposed. When in response one sails out bravely in pursuit of a serious science of human-and-Web intelligence (our H+W), one soon confronts the challenge of rigorously modeling intensionality in many contexts, and

[9] Using Kant's terminology, one can say that humans seem to have access to "a thing-in-itself"—at least as long as the thing is purely abstract.

[10] Some readers will have noted that comparisons can be drawn between our argument and philosophical arguments against so-called Strong AI and machine consciousness and qualia (e.g., see those in Bringsjord 1992), but we refrain from exploring such comparisons here.

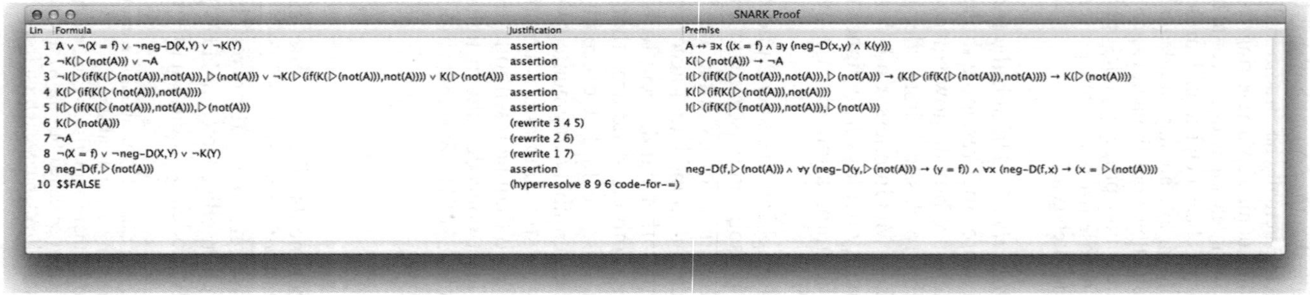

SNARK Proof

Lin	Formula	Justification	Premise
1	A ∨ ¬(X = f) ∨ ¬neg-D(X,Y) ∨ ¬K(Y)	assertion	A ↔ ∃x ((x = f) ∧ ∃y (neg-D(x,y) ∧ K(y)))
2	¬K(▷(not(A))) ∨ ¬A	assertion	K(▷(not(A))) → ¬A
3	¬I(▷(if(K(▷(not(A))),not(A))),▷(not(A))) ∨ ¬K(▷(if(K(▷(not(A))),not(A)))) ∨ K(▷(not(A)))	assertion	I(▷(if(K(▷(not(A))),not(A))),▷(not(A))) → (K(▷(if(K(▷(not(A))),not(A)))) → K(▷(not(A))))
4	K(▷(if(K(▷(not(A))),not(A))))	assertion	K(▷(if(K(▷(not(A))),not(A))))
5	I(▷(if(K(▷(not(A))),not(A))),▷(not(A)))	assertion	I(▷(if(K(▷(not(A))),not(A))),▷(not(A)))
6	K(▷(not(A)))	(rewrite 3 4 5)	
7	¬A	(rewrite 2 6)	
8	¬(X = f) ∨ ¬neg-D(X,Y) ∨ ¬K(Y)	(rewrite 1 7)	
9	neg-D(f,▷(not(A)))	assertion	neg-D(f,▷(not(A))) ∧ ∀y (neg-D(y,▷(not(A))) → (y = f)) ∧ ∀x (neg-D(f,x) → (x = ▷(not(A))))
10	$$FALSE	(hyperresolve 8 9 6 code-for-=)	

FIGURE 9.5. Proof by SNARK of the Knower Paradox

therefore the challenge of modeling many propositional attitudes. We have briefly demonstrated that seemingly sensible proposals for handling even just one such attitude (*knows*) are problematic. We believe the central issue is due to a category confusion: the objects of intensional attitudes such as knowing are propositions, yet all three attempts considered above ignore this brute fact. After all, our Smith may not be a genuine genius because of his brain-to-Web link, but just like us, he certainly knows many propositions. He knows, for example, that the Web houses myriad facts. In knowing this, he doesn't know a string or a number; again, he knows a *proposition*. Yet the attempts briefly canvassed above replace the target of propositional knowledge with, respectively, strings and natural numbers. So long as one fails to acknowledge explicitly that knowing-that is knowing a proposition, and resorts to a "bag of tricks" to squeeze the intensional into the extensional, problems will persist. If they do, not only will we fail to develop a rigorous science of H+W, we won't any time soon even engineer a Simon-like agent capable of approaching genuine intelligence by leveraging the Web, but will instead be forced to content ourselves with only glorified digital turks capable of just fundamentally shallow feats.

Appendix

Figure 9.5 contains the machine proof of the Knower's Paradox proposition proved in the workspace shown in figure 9.4.

Acknowledgments

We are profoundly grateful for support from the John Templeton Foundation that has made possible the research reported on in this chapter. We are also indebted to two anonymous reviewers for insightful comments and objections, and to the editors for sagacious guidance.

References

Anderson, C. Anthony. 1983. "The Paradox of the Knower." *Journal of Philosophy* 80, no. 6:338–55.

Baader, Franz, Diego Calvanese, and Deborah McGuinness (eds.). 2007. *The Description Logic Handbook: Theory, Implementation*. Second edition. Cambridge: Cambridge University Press.

Boolos, G. S., J. P. Burgess, and R. C. Jeffrey. 2003. *Computability and Logic*. Fourth edition. Cambridge: Cambridge University Press.

Boolos, G. S., and R. C. Jeffrey. 1989. *Computability and Logic*. Cambridge: Cambridge University Press.

Bringsjord, Selmer. 1992. *What Robots Can and Can't Be*. Dordrecht: Kluwer.

————. 2001. "Are We Evolved Computers? A Critical Review of Steven Pinker's *How the Mind Works.*" *Philosophical Psychology* 14, no. 2:227–43. (A more detailed version of this paper is available from the author, and is currently available online at: http://www.rpi.edu/~faheyj2/SB/SELPAP/PINKER/pinker.rev2.pdf. Last Accessed April 20, 2012.)

Bringsjord, Selmer, Joshua Taylor, Andrew Shilliday, Micah Clark, and Konstantine Arkoudas. 2008. "Slate: An Argument-Centered Intelligent Assistant to Human Reasoners." In *Proceedings of the 8th International Workshop on Computational Models of Natural Argument (CMNA 8)*, edited by F. Grasso, N. Green, R. Kibble, and C. Reed, 1–10. Patras, Greece: ECAI. URL: http://kryten.mm.rpi.edu/Bringsjord_etal_Slate_cmna_crc_061708.pdf. Last Accessed April 20, 2012.

Cassimatis, Nicholas. 2006. "Cognitive Substrate for Human-Level Intelligence." *AI Magazine* 27, no. 2:71–82.

Descartes, René. 1911. *The Philosophical Works of Descartes.* Volume 1. Translated by Elizabeth S. Haldane and G. R. T. Ross. Cambridge: Cambridge University Press.

Ferrucci, David, Eric Brown, Jennifer Chu-Carroll, James Fan, David Gondek, Aditya A. Kalyanpur, Adam Lally, J. William Murdock, Eric Nyberg, John Prager, Nico Schlaefer, and Chris Welty. 2010. "Building Watson: An Overview of the DeepQA Project." *AI Magazine* 31, no. 3:59–79.

Fikes, Richard, Deborah McGuinness, and Richard Waldinger. 2002. "A First-Order Logic Semantics for Semantic Web Markup Languages." Technical Report, Knowledge Systems, AI Laboratory, Stanford University KSL-02-01.

Goble, Lou, ed. 2001. *The Blackwell Guide to Philosophical Logic.* Oxford: Blackwell.

Inhelder, Barbel, and Jean Piaget. 1958. *The Growth of Logical Thinking from Childhood to Adolescence.* New York: Basic Books.

Russell, Stuart, and Peter Norvig. 2002. *Artificial Intelligence: A Modern Approach.* Upper Saddle River, N.J.: Prentice Hall.

Smith, Peter. 2007. *An Introduction to Gödel's Theorems.* Cambridge: Cambridge University Press.

CHAPTER 10

THE WEB AS A TOOL FOR PROVING

PETROS STEFANEAS AND IOANNIS M. VANDOULAKIS

1. Introduction

With the use of the Web, proving can be practiced as a collaborative activity, involving people with different backgrounds, viewpoints, and research interests. The use of this new medium can cause significant changes in the practice of proving and thereby in our view of proofs. Proofs exist everywhere: in mathematics, in the physical sciences, in computer science, in philosophy, in legal argumentation, and elsewhere. Mathematical proofs, however, are specific only to mathematics.

What is a proof? A preliminary general answer can be that a proof is anything that can convince someone else of the validity of a claim. In different fields there are different definitions or requirements as to what constitutes a proof. In mathematics, a proof establishes the truth of a proposition on the grounds of already established true propositions or axioms; a proposition of which the truth is established is called a theorem. In the physical sciences, a scientific proof can be grounded on experimental data and observations. Philosophical proof is an inference concluded from a series of small, plausible arguments that can typically be considered persuasive. Legal proofs are reached by a jury on the grounds of allowable evidence presented at a trial. Proofs in computing can be programs that prove properties of systems.

Modern mathematical logic has developed a powerful collection of tools for the formal representation of proofs in more than one way. Since the proofs are represented in a formal language, proof theory confines itself only to those propositions which are expressible in a formal language. This was initiated early in the twentieth century by David Hilbert's program (Hilbert 1935), which requires prior formalization of all of mathematics in axiomatic form, so that a proof of the consistency of this axiomatization of mathematics may be examined.

Nevertheless, the traditional approaches to the concept of proof are not adequate for capturing Web-based proving activity, with its new and uncommon features. We claim that these new features can be incorporated in an ideal fashion inside a broader theoretical framework for the understanding of the proving activity initially based on the concept of

Philosophical Engineering: Toward a Philosophy of the Web, First Edition. Edited by Harry Halpin and Alexandre Monnin. Chapters © 2014 The Authors except for Chapters 1, 2, 3, 12, and 13 (all © 2014 John Wiley & Sons, Ltd.). Book compilation © 2014 Blackwell Publishing Ltd and Metaphilosophy LLC. Published 2014 by Blackwell Publishing Ltd.

proof-event introduced by Joseph Goguen (2001) and further developed by us (in Vandoulakis and Stefaneas forthcoming a and b; Stefaneas and Vandoulakis 2011a).

2. The Web as a Tool for Proving

We start our inquiry with an overview of two projects that have used Web-based communication as an essential part of proving. The first project is based on the development of pieces of software, while the other mainly involves crowdsourcing.

Web-based communication for research purposes has been used successfully outside mathematics as well. The Foldit project (http://fold.it/ portal), for instance, is an online video game about protein folding that was developed by the University of Washington's Center for Game Science in collaboration with the Department of Biochemistry as part of an experimental research project. The gamer's highest-scoring solutions were filtered and analyzed by the researchers to identify a structural configuration that could be applied in relevant real-world problems. Although this game is not related to any kind of proving activity per se, it involves crowdsourcing and consideration of the spatial reasoning abilities of the gamers to achieve a better understanding of the structure of natural proteins.

2.1. The Kumo Proof Assistant and the Tatami Project

The Tatami project is a Web-based distributed cooperative software system initiated by Goguen that consists of a proof assistant, called the Kumo system, a generator for documentation Web sites, a database, an equational proof engine, and a communication protocol to maintain the truth of distributed cooperative proofs.

For each proof, Kumo generates a proof Web site (*proofweb*) based on user-provided sketches in a language called Duck and assists with proofs in first-order hidden logic (Goguen 1999, 141). It supports proof debugging, documentation, and explanations (tutorials), and distributed cooperative proving. Parts of a proof can be exchanged among the members of the same group. Web-based communication is considered an essential part of proving.

On the other hand, the understanding of mathematical proofs is facilitated by the Tatami project (Goguen et al. 2000), because it displays them as representations of their "underlying mathematics." By the term "underlying mathematics," Goguen understands not only the tree structure of proofs but also a drama-style structure (in Aristotle's [1997] sense of drama as conflict), a narrative structure (in the spirit of Labov's [1972] and Linde's [1981] ideas), hyperlinks relating material, and image schemas (in Lakoff and Johnson's [1980] sense).

The proofwebs generated by the Kumo proving system are displayed as hypermedia Web sites called *proof pages* or *Tatami pages*, using already available tools like HTML, Java, JavaScript, and a browser, according to the so-called *Tatami conventions*, grounded in narratology (Goguen 2005, 143–44).

Despite its novelty, Kumo had limited impact on automated theorem proving. We argue that Kumo can be considered an early predecessor of more recent projects, such as Polymath and Tricki, despite the fact that there is no evidence that the creators of the latter knew of its existence.

2.3. The Polymath and Tricki Projects

These projects were initiated in 2009 by Timothy Gowers, a Fields Medal–winning Cambridge mathematician, who posed in his blog the question: "Can we have collaborative proofs in mathematics?" (Gowers 2009a). In order to stimulate scholarly reaction, he formulated a mathematical problem in his blog, namely, a special case of the density Hales-Jewett theorem (Hales and Jewett 1963), and invited the entire mathematical community to collaborate openly in finding an alternative proof. The participants in this experiment had to use the comment function of Gowers's blog to suggest ideas, approaches, comments, and pieces of proof. The theorem had already been proved elsewhere, but the proofs were rather long and complicated. A new, "better" proof was required to enhance the understanding of the theorem.

The attempts were developed along two main threads of discourse in the respective blogs of Gowers and Terence Tao—also a Fields Medal winner and a mathematician at the University of California, Los Angeles—following two different approaches. The first approach was predominantly combinatorial, whereas the second was focused on the calculation of bounds of density Hales-Jewett numbers and Moser numbers for low dimensions. After seven weeks of hard work, Gowers announced that the problem was "probably solved" (Nielsen 2009). Both threads of the Polymath project reached a proof, producing at least two new papers that were published under the pseudonym D. H. J. Polymath (2009, 2010a, 2010b).[1]

The positive outcome of the experiment raised serious questions about the advantages of this innovative mathematical practice and its collaborative character. Michael Nielsen uses the term "networked science" to denote the kind of open science that is discovered by new cognitive tools, facilitated by Web tools (Nielsen 2011), whereas Jean Paul Van Bendegem claims that the Polymath project has interesting consequences for the philosophy of mathematics as well (Van Bendegem 2011).

Alongside the Polymath project, in March 2009 Gowers, together with Olof Sisask and Alex Frolkin, launched Tricki, a Wikipedia-style project

---

[1] More details on the Polymath project can be found in Gowers and Nielsen 2009.

of creating a large repository of articles about the mathematical techniques that are useful for various classes of mathematical problem solving (Gowers 2009c; Tao 2009).

As it has been conceived Tricki seems to be a "treasury" of higher-order mathematical thinking. It is a source of specific mathematical techniques illustrated by examples, problem-solving strategies, and methodological hints, as well as personal success stories in mathematical research. It is designed to support the mathematical proving practice. Tricki has all the advantages that the open content on the Web has—that is, it can be reused and shared by the various participants in a mathematical proving activity. Finally, it is the precipitation of the research experience of the mathematical community. Web-based proving amplifies collective creative thinking, whereas Tricki-style projects serve the building up of collective memory on proving strategies.

3. Original Features of Web-Based Proving Activity

Web proving appears to be a novel kind of proving activity with certain original characteristics.

3.1. Change of the Communication Medium

The crucial factor in Web-based proving activity is the use of the Web, in one or way or the other, as both a source of information and a medium of communication. As a source of information, it is a repository of information, ideas, and methods available for eventual use. As a means of communication, the Web does not only offer new possibilities of interpersonal communication, such as e-mail, social networks, and so on; it primarily facilitates the creation of global interest-based communities. This is attained, for example, by means of blogs, which consist of "personal articles" that people post, or wikis, which are electronic documents that people create in collaboration.

The first and foremost feature of the Web that has affected mathematical practice is its *openness*. Contrary to the traditional communication methods, which are one-to-one or one-to-many, the Web-based communication methods have the character of many-to-many. This changes radically the nature of communication: it is no longer communication among the members of a closed community (for instance, the Pythagorean brotherhood) but open, in principle, to everybody—to any interested person or social group.

3.2. Change in Mathematical Practice

The choice of an innovative medium of communication between mathematicians has engendered radical changes in the standard practice of mathematical problem solving. What Gowers actually used was the function of blogs to create interest-based communities, to set up a group of persons interested in research on a particular problem. Since blogs are interactive, Gowers's blog allowed its visitors to leave comments and messages facilitating the exchange and cross-fertilization of ideas; it is this distinctive *interactivity* that enabled the formed group to collaborate on the problem posed and behave like a goal-directed system. This also resulted in unexpected intensification of research work. The speed with which problems can be solved using Web tools cannot be reached by any ordinary collaboration. Gowers has vividly expressed this new fact by saying that "it felt like the difference between driving a car and pushing it" (Rehmeyer 2009).

Interactivity is the cornerstone of Goguen's project as well, in which a *virtual reality* environment is constructed with new genres and metaphors for information visualization and proof representation to facilitate the search for and understanding of proofs over the Web. This is attained by explicating in rigorous mathematical terms—such as sign (or semiotic) systems and semiotic morphisms—the traditional concepts of semiotics, and considering social aspects, such as those that arise in shared worlds.

3.3. Interactivity and Brainstorming

Interactivity enabled the use of a group problem-solving technique known as brainstorming (Osborn 1963), by which a group tries to find a solution for a given problem by gathering a list of spontaneously generated ideas contributed by its members.[2] The method that was actually involved in the Polymath project could be called (asynchronous) *computer-mediated* or *Web-based (group) brainstorming* (Dennis and Valacich 1993), since the blog enabled the creation of a shared list of ideas and amplifying collective intelligence, although no specific brainstorming software was used for this purpose. This was made possible by the rules that the blog administrator defined (Gowers 2009a), which are similar to the basic rules in brainstorming (Osborn 1963): comments had to be short, as easy to understand as possible and not too technical; incomplete and unusual ("stupid," as Gowers put it) ideas were welcomed; technical work had to be postponed until clearly needed; and so on. These rules are designed to stimulate idea generation and increase the overall creativity of the group.

Web-based brainstorming made it possible to take advantage of the complementarity of mathematical capabilities of the contributors.

---

[2] Osborn has postulated that brainstorming is more effective than individual working in generating ideas. Multiple studies have been conducted to test Osborn's thesis, with varied outcomes.

Different specialists know different things; they can speak different (technical) languages; they may share different scientific standards and values; they have different methodologies and styles of inquiry. As Gowers (2009a) explains, "Different people have different characteristics when it comes to research. Some like to throw out ideas, others to criticize them, others to work out details, others to re-explain ideas in a different language, others to formulate different but related problems, others to step back from a big muddle of ideas and fashion some more coherent picture out of them, and so on. A hugely collaborative project would make it possible for people to specialize." When they are gathered together to attack a specific problem, they bring with them their own mathematical cultures and virtuosities. The expertise achieved by such a group cannot be reduced to the sum total of its constituents: it is not just the sum of knowledge of different areas of mathematics that the individuals have.

Subsequent to the brainstorming phase or the collection of contributions, ideas, and proofs of substatements is the phase of categorization and integration. This task is allocated to a member of the group or is undertaken by the blog administrator. Thus the roles of the blog administrator and the integrator are crucial. In the case of Polymath, the blog administrator is an already recognized member of the academic community, defines the problem, is the master of the code of ethics, and monitors the whole process. The integrator acts as proof checker and integrates the different parts of the proof into a unified whole. It is important that Gowers had an undisputably high reputation among the participants.

The Kumo system, on the other hand, lacks the immediate brainstorming functionality. However, it allows the exploration of dead-end strategies, due to its drama-style structural component, which is used to structure the Web sites it generates to display proofs. Informal discussions, possible strategies, failures, and so forth, may be displayed in the Tatami pages, which can be browsed in an order designed by the prover. Kumo lets users do the hard task of proof planning, while machines do the routine work by reducing problems to tractable subproblems for proof engines anywhere on the Internet (Goguen, Mori, and Lin 1997, 24). This facility is not provided in the Polymath project: the comment facility of the blog may support exchange of ideas for proving, but it cannot conduct the act of proving itself.

It is noteworthy that both the Polymath project and the Kumo project are powered by new cognitive tools: the former makes use of socially shared tools, such as blogs, wikis, forums, and so on, and Web-based collaborative brainstorming and argumentation facilitated by the functionalities of the corresponding selected tool; the latter makes use of dynamic modelling tools (Kumo virtual-world proofwebs and [Tatami] proof pages) that also enable automatic proving and proof checking.

3.4. Crowdsourcing and Open-Source/Content Features

It has been argued that Gowers's initiative makes use of "principles similar to those employed in open source programming projects" (Nielsen 2009). However, open-source production is a cooperative activity initiated and voluntarily undertaken by members of the public to prepare essential elements of a product (e.g., the source code for software) aimed at the improvement of an existing product that is freely distributed; the final product has no specific author or inventor.

In Web-based proving, the members of the group involved (e.g., in the Polymath and Kumo projects) contribute to the proof by suggesting ideas and subproofs, which can potentially be integrated into the final proof. However, they do not improve an already existing proof. The final proof, if reached, is original and can have a collective author. Thus, it is the *modular* contribution to proving (the breaking down of a problem into subtasks and subproofs) that resembles the open-source cooperative activity.

Crowdsourcing is a form of activity similar to open sourcing that is initiated by an individual entity (person or company); the work may be undertaken on an individual or a group basis; the final product is owned by the entity that posed the call for a solution to the problem (Brabham 2008). One supposes that this practice makes it possible to expand the size of potential contributors, ensuring deeper insight into the problem.

Gowers's Polymath project resembles a crowdsourcing activity, since Web proving is initiated by the blog administrator and proceeds on a group basis. The final proof, however, cannot be credited to the initiator; it is a kind of collective intellectual property with identifiable individual contributions from every member of the group involved.

3.5. Credibility of Web Proofs

Web-based proof-events raise a very subtle problem: the question of authorship or credibility of a proof.

Traditional proving is considered an individual enterprise of a closed group. The contribution of the different provers in the group is required to be distinct and identifiable. This requirement has raised many controversies about the historical priority of the same or nearly the same discoveries. We would mention here, for instance, the controversy of the discovery of analytical geometry between Descartes and Fermat, of infinitesimal calculus between Newton and Leibnitz, of non-Euclidean geometry between Lobachevsky and Bolyai; there are many other instances. Solving such problems of historical priority is essential for the mathematical community, because it establishes a hierarchy based on a system of common values shared by the community.

In the case of Web-based proving, however, authorship is not a matter of primary importance. The individual character of proving becomes blurred and is displaced by a kind of "massive" (group) character of proving that is shaped as a unified flow of mathematical progress on a certain problem. For the provers involved in Web proving, matters of priority play a subordinate role. Thus the use of new communication media has increased the gap between the self and the Net, which is a particular case of the more general opposition between identity and globalization in the new information age.

This changes radically the structure of the mathematical community. The structure of the community of Web provers is no longer based on the concept of academic hierarchy but is instead based on the concept of "social symmetry," in the sense of equal opportunities presented to the provers for getting involved in the process of mathematical creation, irrespective of their academic status. Nevertheless, mathematical creation does not lose completely its fundamental feature of being a primarily individual enterprise, because the mutual (reciprocal) interest of every prover in the contribution (ideas and subproofs) of another prover does not disappear; on the contrary, it is the cornerstone of this kind of "massive" mathematical activity. Credibility is thus ensured by the publicity surrounding the relevant Web-based proof-events. Individual contribution is recoverable, although it remains unquantifiable and is not currently acceptable for academic credentials.

4. Traditional Concepts of Proof and the Question of Use of New Technologies in Proving

The characteristics of the proving activity described above have raised many questions concerning the nature of proof and the scope and significance of the means used in proving. Foremost among these questions is how technological tools affect creativity and proving efficiency.

The Greek concept of proof from assumptions, as exemplified by *geometrical demonstration* in Euclid's *Elements*, is considered a cardinal breakthrough in comparison to the proofless mathematical practice prevalent in pre-Hellenic civilizations. In the Pythagorean arithmetic tradition, there is no clear concept of proof either. Arithmetical propositions are shown by theoretical *visual* reasoning concerning the possibility of carrying out certain genetic constructions over a domain of concrete objects. These objects are introduced by genetic constructions and represent infinite sequences that are illustrated incompletely by means of a finite suite or configuration (Vandoulakis 2009a, 2010). Euclidean arithmetic is constructed in an *effective* manner as a formal theory of numbers-*arithmoi*, represented by segments, taking for granted the arithmetic of *multitudes* or *pluralities* (Vandoulakis 1998). Geometrical demonstration marks a shift from the genetic-effective style of ancient arithmetical thinking into the geometrical deductive style from assumptions, in the sense of Euclid and other geometers of classical antiquity.

During the fifth century B.C., the search for proof was made pervasive in all spheres of Greek culture. Attempts to devise proofs for some very general statements about the structure of the world, that is, *metaphysical* or *ontological proofs*, appear already in Presocratic philosophy. *Sophistic reasoning*, in which fallacious reasoning is presented as if it were valid proof, is ascribed to the Sophists by Aristotle and Plato. Alongside these developments, theories of truth were advanced that could serve the needs of the relevant mathematical proofs (Vandoulakis 2009b). Furthermore, Aristotle in his *Analytics* made explicit, for the first time, the distinction between *apodictic* and *dialectical* proof.[3] In both cases two persons are assumed to be involved in the process of reasoning: a person who asks a question, or the *inquirer*, and a person who provides an argument for the answer, or the *reasoner*. Accordingly, a proof is viewed as a method of arranging concepts and arguments in order to convince someone else that a statement is valid or true.

The Greek concept of apodictic proof was further refined during the seventeenth and eighteenth centuries as a result of the development of new concepts in the field of calculus. These developments led to a substantial change in the standards of rigor of mathematical proof that were inherited from the ancient Greek mathematicians and the shaping of the concept of *analytic proof*.

At the beginning of the twentieth century, the foundational crisis in mathematics and logic led to the fundamental distinction between the *classical* and the *constructive* concepts of mathematical proof and the elaboration of different systems of mathematics that make use exclusively of either the one or the other concept of proof. David Hilbert (1862–1943) and the Bourbaki group defined the standards for classical proof, while Luitzen Egbertus Jan Brouwer (1881–1966) and his pupil Arend Heyting (1898–1980) determined the standards for constructive proof. This distinction made proof dependent on the *admissible axioms* or *admissible rules of inference* for the proof of mathematical theorems (see Vandoulakis and Stefaneas forthcoming a).

The question of the use of technology in mathematical proving was posed in a dramatic way in 1976, when two mathematicians, Kenneth Ira Appel (born 1932) and Wolfgang Haken (born 1928), presented a computer-generated "proof" of the famous *four-color theorem*, which no human had the means to check so as to be personally convinced of the validity of the argument. The question of whether or not the work of Appel and Harken constitutes a "proof" caused great controversy within the mathematics community. Goguen's and Gowers's projects on the use of Web technology in mathematical proving constitute a new challenge.

[3] Aristotle 1955, *Sophistic Refutations*, I. 165a 19–37; Aristotle 1960, *Topica*, I.i.100a25–30; Aristotle 1960, *Posterior Analytics*, I.i.24a 22–24b 12, xix, 81b 18.

The situation seems to have many analogies to Galileo's decision to use a telescope as a means to explore the sky. Galileo's innovation was not unreservedly welcomed by the community of his day. It met with a mixed reception. There was an enthusiastic embrace of it by some Jesuit astronomers, but many doubts were raised by those who were committed to old medieval standards of astronomical practice. The use of an instrument like a telescope as a powerful technological tool went beyond the standard practice of the times.

The current problem resides, in our view, in the fact that the traditional concepts of mathematical formal proof, as developed throughout history, are inadequate to capture the peculiarities of modern forms of mathematical proof, such as Web-based proofs. The main shortcoming of the traditional concepts is that they do not take into account major communicational and social aspects of the new forms of proving practice. This deficiency can be restored by the introduction of the concept of proof-event.

5. Web Proving as a Particular Type of Proof-Event

The concept of *proof-event* was introduced by Goguen (2001) as a generalized concept incorporating all kinds of proof—mathematical, dialectical, computer, and other proofs:

> A proof event minimally involves a person having the relevant background and interest, and some mediating physical objects, such as spoken words, gestures, hand written formulae, 3D models, printed words, diagrams, or formulae (we exclude private, purely mental proof events . . .). None of these mediating signs can be a "proof" in itself, because it must be interpreted in order to come alive as a proof event; we will call them *proof objects*. Proof interpretation often requires constructing intermediate proof objects and/or clarifying or correcting existing proof objects. The minimal case of a single prover is perhaps the most common, but it is difficult to study, and moreover, groups of two or more provers discussing proofs are surprisingly common (surprising at least to those who are not familiar with the rich social life of mathematicians—for example, there is research showing that mathematicians travel more than most other academics).

An attempt has been undertaken (Vandoulakis and Stefaneas forthcoming b; Stefaneas and Vandoulakis 2011a) to transform this concept into a theoretical framework for the understanding of the mathematical proving activity as an interaction that involves two principal agents—a prover and an interpreter—and requires validation by an appropriate social group. We have distinguished seven essential components of the concept of proof-event and described some types of it. In the present chapter, we claim that Web proving is a proof-event, in particular, a *Web-based proof-event* (Stefaneas and Vandoulakis 2011b).

5.1. Social Component

Proof-events are social events that take place in space and time and involve particular persons that form social groups of experts with particular knowledge and skills.

In the case of Web proving the relevant group is generated by a Web component (blog, wiki, forum, and so forth) run by a host. The social group generated in such a way is principally an open group that bears no internal hierarchical structure, like the traditional academic communities in educational and research institutions, except the host, who enjoys the privileges of administrator. Thus the generated communities are the product of the free choice of their members to join an event that focuses on a problem or topic of their interest.

5.2. Communication Medium

Proof-events are transmitted through certain communication media, such as text (manuscripts, printed or electronic texts, letters, shorthand notes, and so on) written in (ordinary or formal) language or any other semiotic code of communication (signs, formulae, and so on), oral communication (speeches, interviews, and so on), visual (nonverbal) communication (diagrams, movies, Java applets, and so on), and communication through practices.

The expansion of possible communication media by integrating a variety of Web tools entails changes in the structure of proof-events: the generation of proof in Web-based proving essentially relies on the interactive capacity of the underlying Web tools, which is used by visitors to take active part in an ongoing process of creation and collaboration. The result of this process of creation is articulated in the form of modules (ideas, comments, proofs of substatements, and so on) suggested by the members of the group.

5.3. Prover-Interpreter Interaction

Proof-events presuppose at least two agents: a prover, which can be a human or a machine, and an interpreter, who must be only human (a person or group of experts).

In the case of Web proving the prover and the interpreter are separated by space and time, but they are in communication. Moreover, all the members of the relevant group may play both roles interchangeably. Thus, the prover-interpreter interaction becomes more flexible, since a prover can also be an interpreter of a sub-proof-event of another prover, and vice versa. In this way, a prover and an interpreter may often change their roles over time during an ongoing proof-event.

5.4. Interpretation Process

By interpretation we mean the determination of the definition or meaning of the signs that are fixed by the language or semiotic code of communication used for the presentation of an idea or a proof or subproof or what is thought of as a proof. This aspect is essential, since in Web proving specialists of various profiles are involved.

Interpretation is an active process that is greatly facilitated by the interactivity characteristics of the Web tool employed. During the interpretation process, the interpreter may modify an idea or alter a suggested subproof by adding new concepts (definitions) and filling possible gaps in the proof. Nevertheless, the process of interpretation is not explicitly stated when a proof-event is under way.

5.5. Understanding and Validation

A proof is completed when the persons involved in a proof-event, that is, the members of the relevant Web-generated group, conclude that they have understood the proof and agree that a proof is actually given, that is, that the proof is a fact. This, however, is not achieved automatically. As Goguen (2001) has noticed,

> A proof event can have many different outcomes. For a mathematician engaged in proving, the most satisfactory outcome is that all participants agree that "a proof has been given." Other possible outcomes are that most are more or less convinced, but want to see some further details; or they may agree that the result is probably true, but that there are significant gaps in the proof event; or they may agree the result is false; and of course, some participants may be lost or confused. In real provings, outcomes are not always just "true" or "false." Moreover, members of a group need not agree among themselves, in which case there may not be any definite socially negotiated "outcome" at all! Each proof event is unique, a particular negotiation within a particular group, with no guarantee of any particular outcome.

It seems that this process was undertaken/activated during the Web proving of the density Hales-Jewett theorem. After seven weeks of Web collaboration Gowers (2009b) announced that the problem "was solved (probably)":

> Why do I feel so confident that what we have now is right, especially given that another attempt that seemed quite convincing ended up collapsing? Partly because it's got what you want from a correct proof: not just some calculations

that magically manage not to go wrong, but higher-level explanations backed up by fairly easy calculations, a new understanding of other situations where closely analogous arguments definitely work, and so on. And it seems that all the participants share the feeling that the argument is "robust" in the right way. And another pretty persuasive piece of evidence is that Tim Austin has used some of the ideas to produce a new and simpler proof of the recurrence result of Furstenberg and Katznelson from which they deduced DHJ [the density Hales-Jewett theorem].

It is noteworthy that Gowers's confidence relies on two evidences:

a) the understanding of the Web-based proof, which is not based on complicated calculations that were difficult to track; and
b) the fact that some of the ideas eventually had clearly understood applications in other situations.

Hence, the "administrator" of the Web medium (Gowers's blog, in the considered case) plays the role of validator. The authority of the validator is rooted in his or her mathematical achievements outside the specific group defined for the particular problem. The administrator also performs general coordination functions: he or she may synthesize (integrate) disperse modules (subproofs) or disregard (reject) unimportant modules or irrelevant ideas. The process of synthesis, however, is performed at two levels:

• the level of the individual prover (local); and
• the level of the administrator (global).

The administrator ensures in this way the validation of the final proof, which may be verified either internally (by the participants/provers of a proof-event) or externally (by independent reviewers).

In the case of Kumo, however, proofs and proof steps can be understood more easily, because they can be viewed as representations in the corresponding (Tatami) proof pages, which incorporate as well prover-supplied informal explanations, eventual failures, and the motivation behind important proof steps. Thus, the understanding is conducted in a virtual world environment.

5.6. Historical Component

Insofar as proof-events take place in space and time, they have a "history." This history can be recovered and restored in the communication medium employed in the proving activity for research and transmission. Because of this fact, the individual contribution of each prover who participates in a Web-based proof-event is principally identifiable at every stage and time interval.[4]

5.7. Styles

Proof-events generate proofs in different styles. Styles characterize the art of proving in different cultures or schools or by scholars who may differ in their perception of rigor and other views of a metatheoretical character.

In the case of the Polymath project there are no studies on the role of differences in styles of research and communication (computer-mediated brainstorming) of the principally international and intercultural groups of participants in a proof-event on creativity, proof making, and integration.

In the case of proving in Goguen's virtual worlds, style is defined as Blending Principle choice by using tools from algebraic semiotics (Goguen and Harell 2004). Style blends and their related blending algorithms should satisfy optimality principles, which are used to determine what the most appropriate blend for a given situation is (Goguen 2005, 132–35).

6. Impact on the Concept of Proof

We proceed now to discuss the question of how new Web technologies have changed our views on proofs. Is there enough support for the argument that a new concept of proof has appeared? But what is such a proof? The difficulty of answering this question has been highlighted by the Russian mathematical logician Vladimir A. Uspensky (born 1930), who states, "Although the term 'proof' is perhaps the most important in mathematics, it does not have an exact definition . . . a proof is simply an argument which we find so convincing that we are ready to use it to convince others" (Uspensky 1989, 10–11).

However, the conclusiveness of a mathematical proof, to which Uspensky reduces the concept of proof, raises new problems, as highlighted in Wittgenstein's "dialogue with Turing":

> *Turing:* Proofs in mathematics vary from rather formal proofs to this kind of thing, where one just tries to show him how the things happens, and if he is convinced then it is all right.
> *Wittgenstein:* . . . what do you *convince* him of?
> *Wisdom:* Instead of saying that he is convinced, couldn't you say that he accepts it?
> *Wittgenstein:* Yes, that gets rid of one absurdity. (In Diamond 1976, 127)

---

[4] See the Polymath project's timeline: http://michaelnielsen.org/polymath1/index.php?title=Timeline

At the beginning of the twentieth century, we faced the situation that a proof that was acceptable for a classical mathematician was not acceptable for an intuitionist mathematician—for instance, a proof that affirms the abstract existence of mathematical entities without providing a way to find them. Thus the space of communication that had so far been shared by all mathematicians was split into two spheres. Both groups redefined the concept of formal proof to clarify their viewpoint.

Since then very different approaches have been developed to define what a formal proof is. Hilbert's concept of formal proof, for instance, relies on the concepts of *formal system* and *formalization*. Once formalization has been attained, a *(linear) proof* of a well-formed formula A is a finite sequence of well-formed formulas, the last of which is A and each of the intermediate ones is either an axiom of the formal system or can be derived from previous formulas, according to the specified rules of inference of the formal system. In this sense, a proof of a proposition is represented as a configuration of signs, constructed according to certain rules. The attempt by Gerhard Karl Erich Gentzen (1909–1945) to define formal proof relies on a generalization of the form of natural deduction judgment, called *sequent,* and a variation of Hilbert's concept of formal system, called *calculus of sequents* (*Sequenzen-Kalkül*) (Gentzen 1969).

These concepts, however, define a proof as a final (static) product of a predominantly dynamic process of discovering, which nevertheless remains beyond any formal description. Major aspects of the process of discovering a proof, such as the communication and interaction among provers or between provers and interpreters, the varied communication codes they may use, their capacity to understand a proof, their efficacy in convincing other people that a suggested proof is correct, and so on, had no room in academic discussions until recently. The use of new technologies in proving has brought these aspects to the fore in a notable way. Nevertheless, it is the practices used in proving that undergo radical changes, not the concept of proof itself. The validation of the final product of these practices is checked and validated by an appropriate community that shares already established mathematical values and standards of rigor.

Therefore, we cannot overestimate the significance of the use of the new technologies in proving. The Polymath project is rather a specific experiment, not a universally acceptable methodology of mathematical research. The initiator of the project is a leading mathematician, recognized for contributions to mathematics in the traditional way. The Kumo and Tatami projects, on the other hand, have produced very complicated software prototypes to serve automated proving over the Web. Nevertheless, none of these projects, whatever their value and power, has reached the status of a new "paradigm" of scientific research. The projects have not displaced the "old paradigm" of individualistic research and discovery. They simply show that new forms of research methodologies too can be successful.

The concept of a proof-event can be successfully used to describe such innovative forms of problem-centered proving activity, and it serves as a general framework of interpretation for such experiments. It has the advantage of being applicable not only to complete proofs but also to incomplete proofs or parts (modules) of a proof or proof steps. It is a "dialogical" or "dialectical" model for the description of the proving activity, since it involves at least two agents: a prover and an interpreter. It leaves open the formal question of what a proof is and highlights the significance of understanding and conviction in the process of validation of a proof. Finally, proof-events allow any semiotic system to be a means of formulation and communication, and they incorporate within themselves the history of proofs and the integration of (possibly diverse) proving styles.

Acknowledgments

We are grateful to the anonymous referees for their valuable suggestions. Petros Stefaneas dedicates this chapter to the memory of Joseph Goguen, his adviser at Oxford. The chapter was partially supported by Project THALIS, "Algebraic Modeling of Topological and Computational Structures and Applications."

This research has been co-financed by the European Union (European Social Fund—ESF) and Greek national funds through the Operational Program "Education and Lifelong Learning" of the National Strategic Reference Framework (NSRF)—Research Funding Program: THALIS.

References

Aristotle. 1955. *On Sophistical Refutations, On Coming-to-be and Passing Away, On the Cosmos*. Translated by E. S. Forster and D. J. Furley. Loeb Classical Library No. 400. Cambridge, Mass.: Harvard University Press.

———. 1960. *Posterior Analytics, Topica*. Translated by Hugh Tredennick and E. S. Forster. Loeb Classical Library No. 391. Cambridge, Mass.: Harvard University Press.

———. 1997. *Poetics*. Translated by Samuel Henry. New York: Dover.

Brabham, Daren C. 2008. "Crowdsourcing as a Model for Problem Solving: An Introduction and Cases." *Convergence: The International Journal of Research into New Media Technologies* 14, no. 1:75–90.

Dennis, Alan R., and Joseph S. Valacich. 1993. "Computer Brainstorms: More Heads Are Better Than One." *Journal of Applied Psychology* 78, no. 4:531–37.

Diamond, Cora (ed.). 1976. *Wittgenstein's Lectures on the Foundations of Mathematics: Cambridge, 1939*. Ithaca: Cornell University Press.

Gentzen, Gerhard. 1969. *The Collected Papers of Gerhard Gentzen*. Edited by M. E. Szabo. Amsterdam: North-Holland.

Goguen, Joseph Amadee. 1999. "Social and Semiotic Analyses for Theorem Prover User Interface Design." *Formal Aspects of Computing* 11:272–301. (Special Issue on User Interfaces for Theorem Provers.)

———. 2001. "What Is a Proof?" Informal essay. University of California at San Diego, http://cseweb.ucsd.edu/~goguen/papers/proof.html, accessed April 2, 2012.

———. 2005. "Steps Toward a Design Theory for Virtual Worlds." In Maria-Isabel Sanchez-Segura (ed.), *Developing Future Interactive Systems*, 116–52. Hershey, Penn.: Idea Group.

Goguen, Joseph, Kai Lin, Grigore Rosu, Akira Mori, and Bogdan Warinschi. 2000. "An Overview of the Tatami Project." In *Cafe: An Industrial-Strength Algebraic Formal Method*, ed. Kokichi Futatsugi, Tetsuo Tamai, and Ataru Nakagawa, 61–78. Amsterdam: Elsevier Science.

Goguen, J. A., and D. F. Harell. 2004. "Style as a Choice of Blending Principles." In Shlomo Argamon, Shlomo Dubnov, and Julie Jupp (eds.), *Style and Meaning in Language, Art Music and Design: Papers from the 2004 AAAI Symposium, October 21–24, Arlington, Virginia*, 49–56. Menlo Park, Calif.: AAAI Press.

Goguen, Joseph Amadee, Akira Mori, and Kai Lin. 1997. "Algebraic Semiotics, ProofWebs, and Distributed Cooperative Proving." In *Proceedings, User Interfaces for Theorem Provers '97, Sophia Antipolis, 1–9 September 1997*, ed. Yves Bertot, 24–34. Sophia Antipolis, France: INRIA.

Gowers, Timothy William. 2009a. "Is Massively Collaborative Mathematics Possible?" *Gowers' weblog* (2009-01-27), http://gowers.wordpress.com/2009/01/27/is-massively-collaborative-mathematics-possible/, accessed April 2, 2012.

———. 2009b. "Problem solved (probably)." *Gowers' weblog* (2009-03-10), http://gowers.wordpress.com/2009/03/10/problem-solved-probably, accessed April 2, 2012.

———. 2009c. "Tricki now fully live." *Gowers' weblog* (2009-04-16), http://gowers.wordpress.com/2009/04/16/tricki-now-fully-live, accessed April 2, 2012.

Gowers, Timothy, and Michael Nielsen. 2009. "Massively Collaborative Mathematics." *Nature* 461 (October 15): 879–81.

Hales, Alfred, and Robert Jewett. 1963. "Regularity and Positional Games." *Transactions of the American Mathematical Society* 106:222–29.

Hilbert, David. 1935. *Gesammelte Abhandlungen*. Vol. 3. Berlin: Springer.

Labov, William. 1972. "The Transformation of Experience in Narrative Syntax." In *Language in the Inner City*, 354–96. Philadelphia: University of Pennsylvania Press.

Lakoff, George, and Mark Johnson. 1980. *Metaphors We Live By*. Chicago: University of Chicago Press.

Linde, Charlotte. 1981. "The Organization of Discourse." In Timothy Shopen and Joseph M. Williams (eds.), *Style and Variables in English*, 84–114. Cambridge, Mass.: Winthrop.

Nielsen, Michael. 2009. "The Polymath Project: Scope of Participation." *Michael Nielsen's blog* (2009-03-20), http://michaelnielsen.org/blog/the-polymath-project-scope-of-participation/, accessed April 2, 2012.

———. 2011 *Reinventing Discovery: The New Era of Networked Science* (e-book). Princeton: Princeton University Press.

Osborn, Alex. 1963. *Applied Imagination: Principles and Procedures of Creative Problem Solving*. Third edition. New York: Charles Scribner's Sons.

Polymath. 2009. *A New Proof of the Density Hales-Jewett Theorem*. At http://arxiv.org/abs/0910.3926, arXiv:0910.3926v2 [math.CO], accessed April 2, 2012.

———. 2010a. *Deterministic Methods to Find Primes*. At http://arxiv.org/abs/1009.3956, arXiv:1009.3956v3 [math.NT], accessed April 2, 2012.

———. 2010b. *Density Hales-Jewett and Moser Numbers*. At http://arxiv.org/abs/1002.0374, arXiv:1002.0374v2 [math.CO], accessed April 2, 2012.

Rehmeyer, Julie. 2009. "Mathematics by Collaboration." *Science News* (2009-12-08), http://www.sciencenews.org/view/generic/id/50532/title/Mathematics_by_collaboration, accessed April 2, 2012.

Stefaneas, P., and I. M. Vandoulakis. 2011a. "Proofs as Spatio-Temporal Processes." In *14th Congress of Logic, Methodology and Philosophy of Science 2011, Nancy, July 19–26, 2011 (France), Volume of Abstracts*, 131–32.

———. 2011b. "Web as a Tool for Proving." The Second International Symposium on the Web and Philosophy, Co-located at Philosophy and Theory of AI Conference (PT-AI), October 5, 2011, Salonika (Thessaloniki), Greece (abstract).

Tao, Terence. 2009. "Tricki Now Live." *What's New* (20009-04-16), http://terrytao.wordpress.com/2009/04/16/tricki-now-live, accessed April 2, 2012.

Van Bendegem, Jean Paul. 2011. "Mathematics in the Cloud: The Web of Proofs." In *14th Congress of Logic, Methodology and Philosophy of Science 2011, Nancy, July 19–26, 2011 (France),Volume of Abstracts*, 14.

Uspensky, V. A. 1989. *Gödel's Incompleteness Theorem*. Moscow: Mir.

Vandoulakis, I. M. 1998. "Was Euclid's Approach to Arithmetic Axiomatic?" *Oriens-Occidens: Cahiers du Centre d'histoire des Sciences et des philosophies arabes et Médiévales* 2:141–81.

———. 2009a. "Styles of Greek Arithmetic Reasoning." *Study of the History of Mathematics RIMS Kôkyûroku* 1625:12–22.

———. 2009b. "Early Greek Mathematics and Eleatic Philosophy: A Reappraisal of a Controversial Relation." *Abstracts of the Symposium "Mathematical Discoveries and Demonstrations: East and West" for the XXIIIth International Congress of History of Science and Technology, Budapest, Hungary, July 28–August 2, 2009.*

———. 2010. "A Genetic Interpretation of Neo-Pythagorean Arithmetic," *Oriens—Occidens: Cahiers du Centre d'histoire des Sciences et des philosophies arabes et Médiévales* 7:113–54.

Vandoulakis, I. M., and P. Stefaneas. Forthcoming a. "Conceptions of Proof in Mathematics." In V. A. Bazhanov, A. N. Krichevech, and V. A. Shaposhnikov (eds.), *Proof: Proceedings of the Moscow Seminar on Philosophy of Mathematics* [Доказательство: Труды Московского семинара по философии математики / В. А. Под ред, Бажанова, А. Н. Кричевца, В. А. Шапошникова] Moscow. (In English, with summary in Russian.)

———. Forthcoming b. "A Typology of Proof-Events." In S. L. Singh et al. (eds.), *Proceedings of the Symposium in Memory of Professor B. S. Yadav (1931–2010).* Almora, Uttarakhand (India).

CHAPTER 11

VIRTUAL WORLDS AND THEIR CHALLENGE
TO PHILOSOPHY:
UNDERSTANDING THE "INTRAVIRTUAL" AND
THE "EXTRAVIRTUAL"

JOHNNY HARTZ SØRAKER

Introduction

The Internet has changed our lives radically, and probably much more than we realize. For pretty much any concept entertained in the history of philosophy, the Internet has caused myriad conceptual muddles (Moor 1985) and brought with it a dramatic re-ontologization (Floridi 2005) of the world towards increasingly digital, as opposed to analogue, entities, events, and experiences. To illustrate, it should suffice to name the challenges posed by the Web to just some of the most basic concepts in philosophy: the notion of "space" is challenged by the way in which the Web creates new spaces for actions and events to unfold, the notion of "time" is challenged by its increasing disentanglement from physical distance, the notion of agency is challenged by "bots" and other artificial agents, and the notion of mind itself is challenged by the way in which many cognitive functions such as memory become causally intertwined with (and increasingly dependent on) instantly available information online (Clark and Chalmers 1998).

Although the Web has had a profound and probably underappreciated effect on most if not all aspects of our lives, we have certainly seen dramatic changes brought about by earlier technologies as well, including the familiar examples of the telescope, the steam engine, and the pre-Web computer. Many of these changes have also corresponded to changing the agenda of philosophers, as evidenced by the intertwining of philosophy of mind with developments in artificial intelligence. In an attempt to understand how technology can change our relation to the world, numerous philosophers have also tried to develop new theoretical frameworks. Two complementary and influential frameworks have been developed by post-phenomenologists Don Ihde (1990) and Peter-Paul Verbeek (2005), and their approaches try to shed light on different ways in which technology changes the relationship between the subject and its lifeworld. In this chapter, I argue that these frameworks are incapable of dealing with the

Philosophical Engineering: Toward a Philosophy of the Web, First Edition. Edited by Harry Halpin and Alexandre Monnin. Chapters © 2014 The Authors except for Chapters 1, 2, 3, 12, and 13 (all © 2014 John Wiley & Sons, Ltd.). Book compilation © 2014 Blackwell Publishing Ltd and Metaphilosophy LLC. Published 2014 by Blackwell Publishing Ltd.

radical yet subtle way in which the Internet changes our relation to the world and others, and through this analysis attempt to show both the uniqueness of the Internet as a technology and the challenge it poses to some of our most fundamental philosophical notions. Furthermore, I introduce an important distinction between what I refer to as "extravirtual" and "intravirtual," which allows us to better conceptualize actions and events that are made possible by the Internet, and which also sheds light on why it is (or should be) so difficult to arrive at an observer-independent ethics for online behaviour.

In order to make this analysis as clear as possible, it is necessary to narrow the scope somewhat, since "the Web" is an enormously multifaceted concept. Although I believe that the discussion below is relevant to many aspects of the Internet, I focus on real-time interactions in three-dimensional environments, since this is the type of Internet technology where the phenomena that I discuss can be most clearly seen. By "virtual world" I refer to computer-simulated, interactive, multi-user, three-dimensional environments, where users can interact with each other by means of graphical representations of themselves ("avatars"). Although virtual worlds are primarily developed as stand-alone applications that are, strictly speaking, not part of the Web, there is a strong push in the industry towards embedding virtual worlds in Web browsers in order to lower the threshold for new, casual customers. Second Life (*Project Skylight*) and Google (*Lively*) have already experimented with a Web browser implementation, but both projects failed due to lack of standards and browser support. Because of these kinds of problems, the Web3D Consortium was founded with the purpose to "create and encourage market deployment of open, royalty-free standards that enable the communication of real-time 3D across applications, networks, and XML web services."[1] This consortium and recent developments towards 3D capabilities in Microsoft's *Silverlight* and Adobe's *Flash* are strong indicators that virtual worlds will increasingly become part of the Web proper. Although in the remainder of this chapter I refer to "virtual worlds," since this is where the phenomena I am interested in can most clearly be seen at present, it seems likely that the full impact of these phenomena will only become widespread when these virtual worlds can be accessed directly from the Web browser.

Narrowing the scope in this manner allows us to partly bracket the fact that "the Internet" will typically entail a mesh of online and offline practices that can hardly be separated. Although I agree that there is usually no "magic circle" (Huizinga 2003) surrounding the virtual, "virtual worlds" are probably as close as we get to having some kind of "membrane" between real and virtual, in so far as virtual worlds allow

---

[1] "WEB3D Consortium Process Summary and Guidelines." Retrieved March 2, 2012, from http://www.web3d.org/files/documents/Web3D_Process_Summary_and_Guidelines_Apr05.pdf.

us to create and maintain relationships entirely independent from our offline lives. As a first step towards understanding the uniqueness of virtual worlds, and the lessons to be learned in philosophy, I will show how some of the traditional frameworks for making sense of how technology alters our relationship to the world become inapplicable to virtual worlds—and that this inapplicability is in itself revealing of its unique characteristics.

Post-Phenomenology and Technology Relations

In his influential *Technology and the Lifeworld* (1990), Ihde conceptualizes four different ways in which technologies mediate between humans and the world. First, some technologies become *embodied,* meaning that the technology in question alters the way in which we perceive the world without the technology itself being explicitly present; the technology "disappears" into the background when it is being used. Typical examples include glasses, microscopes, and telescopes, which allow us to perceive (parts of) the world we would not otherwise see, without noticing the technologies that make this possible. Second, technologies can form an *alterity* relation, in which we interact with the technology itself as an Other—leaving *the world* more or less in the background. Typical examples of alterity relations include the withdrawal of money from an ATM or interacting with a robot. Third, some technologies form part of a *hermeneutic* relation, in which a part or feature of the world can be read (and in some cases interpreted) by human beings by means of a technology. A standard example given for such a relation is the thermometer, which "hermeneutically delivers" a representation of a particular aspect of the world. Finally, Ihde also introduces *background* relations, where the technology is not perceived directly but becomes "a kind of near-technological environment itself" (Ihde 1990, 108). The most obvious examples include the kinds of technologies that surround us in everyday life, such as technologies for lighting, heating, air conditioning, and so on.

Ihde's relations make intuitive sense for many, if not most, types of traditional technology, but they sometimes come up short when it comes to new and emerging technologies—including virtual worlds, which I will return to shortly. Due to the limits of Ihde's analysis, Verbeek (2008) proposes two other types of technology relations. First, in some cases, the relation between human and technology is much closer than that of Ihde's "embodiment" relations. For instance, not only do neural implants and other bionic technologies become embodied in the sense of not being directly perceived, the technology "physically alters the human" (Verbeek 2008, 391). Verbeek dubs this the "cyborg relation." Another technology relation introduced by Verbeek is the "composite relation," in which technologies not only represent a phenomenon in the world but *construct*

TABLE 11.1. Technology relations according to Ihde and Verbeek

Technology relation	Schematization	Examples	Author
Embodiment relation	(Human – technology) → world	Glasses	Ihde
Hermeneutic relation	Human → (technology – world)	Thermometer	
Alterity relation	Human → technology (– world)	ATM	
Background relation	Human (– technology – world)	Air conditioner	
Cyborg relation	(Human/technology) → world	Neural implant	Verbeek
Composite relation	Human → (technology → world)	Radio telescope	

reality instead. For instance, a radio telescope generates an image comprehensible by humans on the basis of detecting radiation that is invisible to the eye. This is not merely a form of bringing what is far away closer or enlarging microscopic entities; these technologies construct that which cannot be perceived by humans—and sometimes generate a representation in an entirely different modality as well, for instance by making sounds visible.

All of the technology relations outlined above can be illustrated in the form of arrows and dashes. In Ihde's original formulation, the dashes constitute what he refers to as *enigma* positions (Ihde 1990, 86–87).[2] These enigma positions are points where the relation may break down in one way or another—or simply the point where the technologies themselves may malfunction. For instance, if glasses break, they no longer form an embodiment relation between the human and the world, and if a thermometer stops functioning, it no longer forms a hermeneutic relation between the human and the world. We can illustrate Ihde and Verbeek's technology relations as in table 11.1.

The combined conceptualizations of Ihde and Verbeek help explain a range of different technologies, but can they shed light on the ways in which the Web—and virtual worlds in particular—change the relationship between subjects, and between subjects and reality?

Applying Technology Relations to Virtual Worlds

When we try to apply these relations to virtual worlds and entities, we quickly find ourselves in trouble—but these problems are in themselves interesting because they reveal part of what is so unique about virtual worlds. One initial problem is that the distinction between "technology"

---

[2] Verbeek removes Ihde's dashes in his two additional relations. In the cyborg relation, this makes sense in a somewhat disturbing manner, since a breakdown of the technology will destroy not only the relation but the human-technology hybrid itself. There seems, however, as if there should be an enigma position between technology and world in the composite relation.

and "world" featured in Ihde and Verbeek's relations becomes complicated when we talk about virtual worlds. We might propose to have "world" refer exclusively to the *physical* world, but this leads to a couple of peculiar problems. First, since we use technologies (computers and peripherals) to access virtual worlds, are these technologies part of the mediation between the human being and the virtual world? Second, how do we conceptualize technologies *within* the virtual world itself? If we look at the examples given above, it seems that all of Ihde's relations can be applied to virtual entities *within* virtual worlds as well: there are virtual glasses, virtual meters of various sorts, virtual ATMs, and virtual lighting. Are these entities part of the virtual-world-as-technology, or are they themselves technologies within another technology? It could be argued that we do not directly interact with these technologies, but reducing a virtual world to our interaction with the computer and peripherals hardly makes sense.

To make matters even worse, the virtual world can be seen as a form of hermeneutic technology as well, because it does mediate the physical world—in particular, the actions of *other* human beings. To illustrate this, take a regular computer game that does not allow for online multiplayer. Such games can be considered as a form of *alterity* relation, because we interact with the technology while the world is more or less in brackets. With virtual worlds (i.e., when there is an online multiplayer element), the world is no longer bracketed because we communicate with actual human beings through the virtual world/technology. This, in turn, is seamlessly combined with all kinds of *non-human* "alters" within the virtual world. Finally, as if the notions of "technology" and "world" are not difficult enough to place in these relations, the concept of "human" is also complicated by the fact that the interactions are carried out *as if* done by a representation of the human (an "avatar"), and done from the standpoint of that avatar's indexical location within the world/technology.

In short, with virtual worlds the relata in the technology relations of Ihde and Verbeek become ambiguous. "Human" may refer to the actual human or to the avatar representation. "Technology" may refer to the user's computer and peripherals, the computer simulation and databases that underpin the virtual worlds, the virtual world itself (as experienced), and/or virtual entities within virtual worlds. Finally, "world" can refer to the actual or the virtual world.

At first glance, it may seem that Verbeek's composite relation is a promising way of conceptualizing virtual worlds as essentially *constructing* reality. This is partly correct if we regard the virtual world as mediating computer states (encoded strings of binary digits) as unobservable aspects of the physical world. However, this fails to capture the relation between two humans interacting with each other through a virtual world. Although it may be technically correct to describe virtual worlds as constructing that which cannot be experienced as such (i.e., the underlying computer states), this misses the *experience* of being immersed in a virtual world and of communicating with another human being by way of avatars—and losing the subjective experience seems particularly unfortunate from a (post-)phenomenological point of view.

The considerations above are not intended merely as a criticism of Ihde and Verbeek. They simply show that virtual worlds and entities probably cannot be conceptualized in the same manner as many, if not most, other technologies. Virtual worlds are both worlds and technologies, the computer simulation is both the underpinning of the virtual world and the means of mediation, entities within virtual worlds can be regarded as technologies themselves, and although virtual worlds mediate the physical world and other human beings, they *also* construct reality. All of this complexity shows how unique virtual worlds are—and how difficult it is to conceptualize the relations between humans, virtual worlds, and the physical world. Thus, we need to simplify and focus on a few, particularly important ways in which virtual worlds are related to other human beings and the physical world. I will do this by way of the terms "extravirtual" and "intravirtual" and will then show how this distinction illustrates the uniqueness of virtual worlds.

Extravirtual and Intravirtual Consequences

One of the reasons why virtual worlds and entities are often considered mere play without any importance for our real lives stems from the belief that virtual entities and events have no effects in the physical world. In other words, there can be no direct *physical* harm coming from online communication, and thus any law or moral principle (implicitly) derived from the harm principle will be inapplicable. This is true if we talk about virtual entities qua virtual, but it quickly becomes more complicated when we take into consideration that any change in the virtual world *necessarily* corresponds to changes in physical reality—that is, changes in the physical computer running the simulation. Thus, in making claims about the effects of virtual worlds on the actual world, we need to distinguish between the virtual entity as such and the causes of its existence.[3] We must not forget that virtual entities are not mere products of the mind or illusions; they are generated and made intersubjectively available by a computer according to a comprehensive set of regulative principles.

This gives rise to a peculiar feature of virtual worlds and entities, which is that the underlying computer simulation can produce effects both within the virtual world and outside it—and this effect can be

---

[3] Another way to put it would be to distinguish between different levels of abstraction (cf. Floridi 1999). In this terminology, what I refer to as intravirtual would be the virtual level of abstraction, and extravirtual would be anything external to the virtual level, such as the computational, physical, or mental level of abstraction.

caused by the same event. To illustrate this point, imagine a virtual world in which I throw a virtual rock that hits a virtual window. The virtual rock qua virtual does not have any mechanico-causal connection to the physical world, but the simulation is generated by a computer, which has both a physical existence and the capacity for causing even dramatic changes in the physical world. Consider, moreover, that breaking the virtual window triggers a certain computer state, which in turn triggers the detonation of a physical bomb. Does this mean that it was the virtual rock that caused the physical explosion? What happens is that the virtual event has two very different kinds of effects, what I refer to as *intravirtual* and *extravirtual* effects. By moving my physical body (extravirtual) in such a way that I throw a virtual rock (intravirtual), I am causing a change in the state of the computer running the simulation (extravirtual). This change of state, in turn, can further produce both intravirtual and extravirtual effects—respectively, the breaking of the virtual window (intravirtual) and the detonation of the bomb (extravirtual). The intravirtual effects are the effects I experience as being part of the virtual world, such as seeing the virtual window being shattered and hearing the corresponding sounds. We can describe these as events that are *congruent* with the virtual world as a whole. In technological terms, these correspond to particular visual stimuli presented through the monitor and sounds emitted through speakers or headphones—or if we speculate into the future of virtual worlds, the perceptual feedback from a head-mounted display and tactile stimuli from a body suit. The important point here is that although the virtual world as such is observer dependent, the computer states that underpin the virtual world and give rise to having a shared experience are themselves observer independent. To get a better understanding of this, it is helpful to employ John Searle's terminology concerning intentionality and the conditions of satisfaction (Searle 1995, 2001).

Intravirtual Versus Extravirtual Conditions of Satisfaction

According to Searle, although beliefs and desires have a similar structure in virtue of being intentional states, their relation to the world falls into two diametrically opposite categories. Beliefs are true or false, whereas desires, intentions, hopes, and so on, are satisfied or frustrated. Searle's way of describing the difference is that in both cases we are talking about "conditions of satisfaction," but they have different "directions of fit." What this means is that for a *belief* to be satisfied, the belief must "fit" the world; it is the belief that must be adjusted so as to be in accordance with the world. For instance, my belief that it is raining is satisfied (i.e., true) if/when it is in fact raining. If it rains but I do not believe it is raining, it is the belief that needs to change so as to be in accordance with the world. For a *desire* (or similar intentional states) to be satisfied, however, it is not the desire that needs to change so as to fit the world, it is the world that must be adjusted so as to be in accordance with (fit) the desire. My desire that it should rain is satisfied if and only if the world changes so as to fit the desire. Thus, beliefs and desires have opposite directions of fit. In Searle's terminology, beliefs have a mind-to-world direction of fit, whereas desires have a world-to-mind direction of fit.[4] One important implication of this is that the rationality of our beliefs and desires will be determined by where the conditions of satisfaction lie. It is perfectly rational for me to believe that Hamlet is the prince of Denmark as long as I restrict its condition of satisfaction to the works of Shakespeare, but it would be entirely irrational if the conditions of satisfaction lie in the actual world.

With this background, we can better clarify problems related to confusing the intravirtual and the extravirtual. First, some problems occur due to our being perfectly aware of the intravirtual and extravirtual consequences of our actions—but the two are in conflict with each other. Consider a man who has the desire to have a virtual relationship with a woman. This desire is imprecise unless it is specified where the condition of satisfaction lies. If the man desires to have a virtual relationship with someone who is a woman (only) in the virtual world, then the condition of satisfaction is intravirtual—and the satisfaction of the desire depends on whether the state of affairs in the virtual world comes to fit his desires. If, however, the man desires a virtual relationship with someone who is a woman in the actual world, then the condition of satisfaction is extravirtual—and the satisfaction of the desire depends on whether or not the state of affairs in the actual world comes to fit his belief. Notice furthermore that whether or not the steps he takes towards satisfying his desires are rational or not will be determined by where the condition of satisfaction lies. This becomes more complex when the extravirtual and the intravirtual become contradictory—and even more so when dealing with emotionally laden activities, as relationships and corresponding activities tend to be.

The distinction between intravirtual and extravirtual also becomes problematic when we mistakenly place the conditions of satisfaction intravirtually, not realizing that they are in fact extravirtual, and this is part of what makes judgments regarding culpability in virtual worlds so difficult. For instance, there have been instances of people who have sued various service providers for not making the extravirtual consequences of their

---

[4] Searle's terminology can become messy at times, especially when he substitutes these terms for, respectively, the "upward" and "downward" directions of fit. I have found that the easiest way to remember the difference between mind-to-world and world-to-mind is to think of the former as "mind-must-conform-to-world" (i.e., beliefs must conform to, or become the same as, the state of affairs in the world in order to be true/satisfied) and of the latter as "world-must-conform-to-mind" (i.e., the state of affairs in the world must conform to, or become the same as, the desire if it is to be satisfied).

actions clear enough, including instances of bankruptcy due to trading what were believed to be virtual stocks—when in fact the "virtual" stocks were as real as it gets. That is, the "virtual" purchases did not only count as the purchase of stocks within the content of a virtual bank (intravirtually), they counted also as the purchase of stocks in the actual world (extravirtually). In this case, it was not the action that was irrational, since the buyer had a desire to buy virtual stocks—and his actions, for all he knew, satisfied this desire *intravirtually*. His belief was irrational, or false, only because he failed to acknowledge that his virtual acts had extravirtual consequences.

The potential conflict between extravirtual and intravirtual conditions of satisfaction points to another peculiar feature of virtual experiences. If we return to the example above, the person seeking a relationship with a virtual woman can wilfully believe that the person behind the avatar is not a man, disregarding any evidence to the contrary, and therefore engage in the perfectly rational and satisfiable desire to have a virtual relationship with a female. Since a desire often has conflicting intravirtual and extravirtual conditions of satisfaction, it is a common strategy among members of virtual worlds to wilfully adopt certain beliefs and avoid evidence to the contrary, precisely in order to make the actions rational *intravirtually*. This is sometimes referred to as willing suspension of disbelief. In Searle's terminology, we can describe suspension of disbelief more precisely as wilfully placing the conditions of satisfaction for the belief within the virtual world alone. For instance, the belief that one has a relationship with a woman is satisfied (i.e., true) when the intravirtual states of affairs fit the belief, regardless of extravirtual evidence to the contrary.

Although wilfully placing the conditions of satisfaction in the virtual world can be a straightforward way to avoid a conflict between intravirtual and extravirtual conditions, it can also give rise to a number of ethically problematic scenarios. In the infamous case of the first virtual rape (Dibbell 2007), the perpetrator took control over another user's avatar and forced it to commit extreme sexual and violent acts that the user would never have consented to. Some, including the perpetrator, would see the notion of virtual rape as entirely misplaced because, after all, the virtual entities involved had no physical properties, hence little to do with the physical aspects of rape. The *extravirtual* consequences, however, do arguably have something in common with rape, such as feelings of shame, loss of autonomy, and a sense of degradation (Søraker 2010). The bystanders' response to the incident ranged from rage towards the perpetrator to annoyance with the victim, and these responses correspond to the complex web of conflicting conditions of satisfaction. That is, our judgment of the severity of an event depends on whether we (implicitly) judge someone's intention as aiming for intravirtual or extravirtual conditions of satisfaction—for instance, whether we judge the virtual rapist as trying to hurt the avatar or the person controlling the avatar. The intravirtual consequences of the virtual rapist's action consisted in nothing but a public, textual description of the actions that the victim engaged in, all of which were beyond her control.[5]

What is important is that the *extravirtual* effects of these intravirtual changes are largely determined by the user's mental states. In Dibbell's case, the victim "was surprised, to find herself in tears—a real-life [extravirtual] fact that should suffice to prove that the [intravirtual] words' emotional content was no mere fiction. . . . Murderous rage and eyeball-rolling annoyance, was a curious amalgam that *neither the RL [extravirtual] nor the VR [intravirtual] facts alone* can quite account for (Dibbell 2007, 15–16; my inserts in brackets). The important point here is that the victim's reaction was largely determined by how emotionally invested she was in her avatar and in the community. The reason why many will have a hard time understanding her reaction ("it's just a game!") stems from the fact that we all have different mental states towards such phenomena. Someone with a casual relationship to her avatar may just as well have found the incident amusing, and we would probably never have heard the story. This is precisely what makes ethics so difficult online: the "invisible" user behind the nick or avatar that we communicate with comes with a set of mental states that determine the extravirtual effects of the intravirtual state of affairs. With traditional, non-virtual examples of inflicted harm, we do not need to ask ourselves whether the subject has a casual or intimate relationship with her own body, but in virtual worlds this makes all the difference.

We are now in a position to pinpoint the radical difference between virtual worlds and other kinds of mediating technologies. When you communicate with somebody on the phone, the mediation can create a sense of remoteness and lack of intimacy that allows us to say things we would not say face to face. The same thing can be said about the extravirtual consequences of virtual communication, but then the remoteness of the extravirtual is not the only thing that might affect your behaviour. In virtual worlds, you get not only the remoteness of the extravirtual but also the immediacy of the intravirtual. For instance, you might say things to an avatar you would not say face to face to an actual person, not only due to being pseudonymous and remote from the *extravirtual* person but also due to the appearance and reactions of the *intravirtual* avatar. That is, when interacting with an avatar, you interact with both the avatar (intravirtual) and the person controlling the avatar (extravirtual). The appearance of the avatar might lead you to interact in a particular way (often

---

[5] The "virtual rape" took place in a text-based virtual world, and the rapist used a script referred to as a "voodoo doll," which meant that he could control the actions of another user. Making the victim engage in sexually deviant and violent actions against her will is, in short, the reason why it has been referred to as a virtual rape.

determined by how you would interact with a similar type of person in the actual world) that could be completely different from how you would interact with the extravirtual person behind the avatar. Thus, the combination of intravirtual and extravirtual gives rise to a unique form of mediation that cannot be found outside virtual worlds. Compare this with the mediation by phone, where it makes little sense to speak of the "intraphonal" aspect of your phone conversation. Thus, when it comes to virtual worlds, Jean Baudrillard is fundamentally mistaken when claiming that when we communicate through computers "the Other . . . is never really aimed at—crossing the screen evokes the crossing of the mirror" (in *Xerox and Infinity*, quoted in Springer 1991, 313). When communicating with (or through) avatars, we do (and should) often have the extravirtual human being in mind. If we do not, we are likely to forget that our virtual acts are perceived by another person—and that they can cause real and strong emotions in that person. All of this becomes even more complex if we are not certain whether the intravirtual entity that we interact with has an extravirtual counterpart or not, which might happen when a virtual world has no clear demarcation between artificial agents (bots) and human representations.

In normative terms, all of this entails that we often *should* remind ourselves of the extravirtual Other, in order to recognize that our virtual acts have potentially dramatic extravirtual consequences. But the fact that the extravirtual consequences are often veiled entails that it becomes difficult to arrive at any kind of observer-independent normative guidelines—at least not beyond some version of a precautionary principle to the effect that, when in doubt, we should always take *potential* extravirtual consequences into account.

The Consequences for Philosophy

Part of the purpose of the discussion above is to illustrate the uniqueness of certain phenomena on the Internet, and how this uniqueness poses problems for traditional philosophical frameworks. I have argued that Ihde's notion of technology relations cannot readily be applied to virtual worlds, and that even Verbeek's explicit attempt to account for new and emerging technologies in a similar fashion also falls short. This illustrates that one of the challenges posed to philosophy lies in the inapplicability of traditional theories and frameworks.

Furthermore, one of the reasons for this inapplicability is itself a challenge to philosophy, to ethics in particular. For any consequentialist type of ethics, the distinction between intravirtual and extravirtual places an extra burden on our ability to predict consequences, not only pragmatically but intrinsically. That is, the very nature of virtual worlds, and most forms of online communication, leaves us with few if any clues as to what the extravirtual consequences of our actions may be. This phenomenon also makes it difficult to judge virtual acts as morally right or wrong. In the case of the virtual rape, this would have amounted to harmless fun (and we would never have heard the story) if there were no extravirtual consequences—extravirtual consequences that are entirely determined by the subjective attitudes of the "victim." The monstrosity of making a similar claim with regard to actual rape—that its severity depends on the victim's attitude to her body—illustrates the vast difference, and the challenges we face when applying ethical theory to virtual worlds. All of this shows that the Internet, and virtual worlds in particular, require entirely new theoretical frameworks—and it is the task of philosophers to critically examine this uniqueness so that we do justice to these phenomena. To be fair, there have been some such efforts recently, most notably the philosophy and ethics of information as primarily developed by Luciano Floridi (1999, 2011), but there is still much work to be done.

Conclusion

I have argued above, using the post-phenomenological notion of technology relation as a starting point, that the Internet, and virtual worlds in particular, come with a set of unique characteristics that leave our traditional frameworks inapplicable. On a more constructive note, I have also tried to show how we can better understand many of these phenomena by introducing new distinctions and frameworks, such as the importance of being clear about the intravirtual and extravirtual consequences of our actions, and the corresponding placement of conditions of satisfaction. This in itself shows how unique virtual worlds are, and the distinctions I have suggested will I hope illustrate the importance of rethinking traditional philosophical concepts and frameworks.

To further illustrate the challenge for philosophy, I have argued that the distinction between intravirtual and extravirtual gives rise to important consequences for judging actions as rational or irrational, and as morally right or wrong. With virtual worlds, these judgments must necessarily be observer dependent in a manner very different from more ordinary phenomena—and in a manner that causes concern for anyone who wishes to address these problems by means of applying traditional philosophical theories. The cause of this lies in the fact that virtual worlds qua virtual are observer dependent (ontologically subjective), yet they are grounded in (or made possible by) observer-independent states of a physical computer—a physical computer that in turn is capable of producing both intravirtual and extravirtual states of affairs. This is like few if any other phenomena, at least none of comparable impact on our lives, and thus requires us to radically rethink our philosophical concepts and frameworks. One hopes this chapter has been a step in this direction.

References

Clark, Andy, and David J. Chalmers. 1998. "The Extended Mind." *Analysis* 58:10–23.

Dibbell, Julian. 2007. *My Tine Life: Crime and Passion in a Virtual World.* Retrieved January 1, 2007, from http://www.lulu.com/content/1070691.

Floridi, Luciano. 1999. "Information Ethics: On the Philosophical Foundation of Computer Ethics." *Ethics and Information Technology* 1, no. 1:33–52.

———. 2005. "The Ontological Interpretation of Informational Privacy." *Ethics and Information Technology* 7, no. 4:185–200.

———. 2011. *The Philosophy of Information.* Oxford: Oxford University Press.

Huizinga, Johan. 2003. *Homo Ludens: A Study of the Play-Element in Culture.* London: Routledge.

Ihde, Don. 1990. *Technology and the Lifeworld.* Bloomington: Indiana University Press.

Moor, Jim. 1985. "What Is Computer Ethics?" In *Computers and Ethics*, edited by Terrell Ward Bynum, 266–75. Oxford: Blackwell.

Searle, John. 1995. *The Construction of Social Reality.* New York: Free Press.

———. 2001. *Rationality in Action: The Jean Nicod Lectures.* Cambridge, Mass.: MIT Press.

Søraker, Johnny Hartz. 2010. "The Neglect of Reason: A Plea for Rationalist Accounts of the Effects of Virtual Violence." In *Emerging Ethical Issues of Life in Virtual Worlds*, edited by Charles Wankel and Shaun K. Malleck, 15–32. Charlotte, N.C.: Information Age Publishing.

Springer, Claudia. 1991. "The Pleasure of the Interface." *Screen* 32, no. 3:303–23.

Verbeek, Peter-Paul. 2005. *What Things Do: Philosophical Reflections on Technology, Agency, and Design.* University Park: Pennsylvania State University Press.

———. 2008. "Cyborg Intentionality: Rethinking the Phenomenology of Human-Technology Relations." *Phenomenology and the Cognitive Sciences* 7, no. 3:387–95.

CHAPTER 12

INTERVIEW WITH TIM BERNERS-LEE

HARRY HALPIN AND ALEXANDRE MONNIN

Harry Halpin: How did the idea of philosophical engineering come about?[1]

Tim Berners-Lee: The phrase came about when we were originally discussing the idea of Web Science,[2] and I was tickled by the fact that when you study and take exams in physics at Oxford, formally the subject is actually not physics but experimental philosophy. I thought that was quite an interesting way of thinking about physics, a kind of philosophy that one does by "dropping things and seeing if they continue to drop"—in other words, "thinking about the stuff you do by dropping things." Then this came up again when trying to explain to people that when we design Web protocols, we actually get a chance to define and create the way a new world works. It struck me what we ended up calling "Web Science" could have been called "philosophical engineering," because effectively when you create a protocol you get the right to "play God" and define what words mean. You can define a philosophy, define a new world. So when people use your system—when they run the protocol—to a certain extent they have to leave their previous philosophy at the door. They have to join in and agree they will work with your system. So you can build systems—worlds—which have different properties. That's exciting, and a source of responsibility as well.

[1] This is a transcript of an interview with Tim Berners-Lee done in Lyon by Harry Halpin and Alexandre Monnin in November 2010, edited for publication and published with permission of Tim Berners-Lee. We have added further comments in footnotes to explain some of the technical terms and background used in this interview by Tim Berners-Lee, who is widely acclaimed as the "inventor of the Web" because he wrote many of the fundamental protocols and created the original prototypes. The first use of term "philosophical engineering" by Berners-Lee in a public forum was this quote: "We are not analyzing a world, we are building it. We are not experimental philosophers, we are philosophical engineers" (Berners-Lee 2003).
[2] Web Science is defined as "a research agenda that targets the Web as a primary focus of attention" (Berners-Lee et al. 2006).

Philosophical Engineering: Toward a Philosophy of the Web, First Edition. Edited by Harry Halpin and Alexandre Monnin. Chapters © 2014 The Authors except for Chapters 1, 2, 3, 12, and 13 (all © 2014 John Wiley & Sons, Ltd.). Book compilation © 2014 Blackwell Publishing Ltd and Metaphilosophy LLC. Published 2014 by Blackwell Publishing Ltd.

Harry Halpin: Would you consider the creation of Web standards to be an act of philosophy in progress?

Tim Berners-Lee: Certainly, when people write a specification, they argue about what words mean until everyone assumes that they mean in some sense the same thing. When the concepts in different people's brains have been sufficiently well aligned and there have been enough connections between the concepts, this is written down in a language that people feel comfortable with and that they share. You can, if you want to, philosophically argue that a word is in fact ambiguous, but nobody bothers. Understand that when you play the game [of Web protocols] you're not going to argue about that. For example, you're not going to pay a bill online, and then afterwards come back and say "Well, I sent some HTTP[3] headers off, but because they're just HTTP headers, they don't actually mean anything." As a spammer said once, "It's just a form field, I can put whatever I like there, it doesn't have to be the person sending the e-mail." But it does if you're playing the game! I think one of the things we're missing is the relationship between "the law of the land" and protocols. It should be easier to establish that when someone disobeys a protocol they've broken a kind of law via a straightforward path.

Harry Halpin: One of the most important aspects of natural language is that it's composed of words. In contrast, the Web is a space of URIs.[4] How is it that URIs and their meaning differs from other possible systems like natural language? What is special about URIs?

Tim Berners-Lee: There are many URI schemes, but one thing that is nifty about HTTP URIs is that they have domain names in them. So they're hierarchical, and a domain is something that one can own.[5] In the way the protocol works, the owner of the domain has the right to say—and the obligation to say on the Semantic Web!—what the things in that domain mean. It's not a question of philosophical discussions

---

[3] HTTP is the HyperText Transfer Protocol (Berners-Lee, Fielding, and Frystyk 1996), the primary protocol defined for use with the Web to deliver web pages through the Internet, although the protocol is now being used for many other applications.

[4] URIs (Uniform Resource Identifiers) are identifiers such as <http://www.example.org/page>, and are defined as an Internet Engineering Task Force Internet standard in various versions (Berners-Lee 1994; Berners-Lee, Fielding, and Masinter 2005). They have also been called Uniform Resource Locations (URLs) due to the debate over whether they "located" or "identified" resources (Berners-Lee, Fielding, and McCahill 1994).

[5] The domain name system refers to the "domain" in a URI. For example, in the URI, the domain is "example.org." The ownership rights of domain names can be purchased via domain registrars, who lease domains.

between third parties. If there's a dispute about what a URI stands for, then the way the protocol works is that you go to the person who owns the domain name, who typically delegates it to someone else, who has in turn designed an ontology that they store on a Web server. The great thing about the Web is that you can look up the HTTP URI in real time to get some machine-readable information about what it means straightaway.

> Alexandre Monnin: Regarding names and URIs, a URI is not precisely a philosophical concept, it's an artifact. So you can own a URI, while you cannot own a philosophical name. The difference is entirely in this respect.

Tim Berners-Lee: For your definition of a philosophical name, you cannot own it. Maybe in your world—in your philosophy—you don't deal with names that are owned, but in the world we're talking about, names are owned. Some people have a philosophy where they find it useful to think of a name as just a function of use, not of definition. Other people like lawyers work in worlds where the model is that there is a legal definition of a term.[6] While meaning is use, use can be according to definition. So there are models, and now we're adding another one, in which meaning is defined by the owner of a name.

> Harry Halpin: Wasn't it controversial that when the Web was first starting that everything could be named with a URI?

Tim Berners-Lee: At the IETF[7] certainly there was resistance. I originally called these things "Universal Document Identifiers" (UDIs) even before we started using them for concepts.[8] The IETF were a bit put off, thinking it was too much hubris to call them "universal." Now I realize that I should have held firm and said "but they are," as any alternative system of naming you can make out there, I can map it to the character set we use in URIs and I can invent a new scheme for it. So we can map any

---

[6] This is a reference to various debates over the "meaning" of URIs. Berners-Lee is likely referring to the defense of "meaning as use" by Yorick Wilks (Wilks 2008), Pat Hayes (Hayes and Halpin 2008), and others.

[7] The IETF is the Internet Engineering Task Force, the primary standardization body of the Internet since its inception. Tim Berners-Lee originally took the Web's primary standards such as HTML, HTTP, and URIs to the IETF, and the Web is considered only one of many possible applications that can run on top of the Internet. He later launched a Web standards body, the World Wide Web Consortium (W3C) in 1994.

[8] A draft called "Universal Document Identifiers" was announced in February 1992. See the message of Berners-Lee to the www-talk mailing list: http://lists.w3.org/Archives/Public/www-talk/1992JanFeb/0024.html.

scheme to URIs. We'd already mapped Gopher, FTP, and these sorts of things.[9] Now, we've got HTTP and there will be lots of other schemes. So in a sense URIs are universal, as we're saying anything—any name that you come across—can be mapped into this space. So yes, there was a lot of pushback against that, and hence the "uniform" rather than "universal" in URIs.[10]

> Alexandre Monnin: Given the origins of philosophical engineering and Web Science, don't you think that Web Science is doing two things? The Web is an artifact, we produce it, we implement it, and as you said, we decide what the protocol means and how it should be used. On the other hand, Web Science is a science, so we make discoveries and we are also surprised by our own creation.

Tim Berners-Lee: The Web Science cycle[11] starts off with idea that the design of the Web is not just the design of one thing but the design of two things: a social and technological protocol. For example, in e-mail, there's a general technological protocol like SMTP[12] and there's a social protocol. In e-mail, the social protocol that states that everyone involved is ready to run a machine that has the space to store e-mail messages while they are en route to their destination, that people will send e-mail to each other on perfectly reasonable topics, and that people will read e-mail that they receive. There's that social piece of e-mail, but then technically e-mail is actually pushed around with SMTP and pulled off with IMAP, and those pieces then together form a system.[13] It's a microscopic system that explains how one person sends another person an e-mail through a finite number of hops, but then you get the effects of scale. So the engineering of Web Science is not like building a mousetrap. You design a microscopic system, but what you're interested in is the macroscopic phenomena that emerge. When you do the science—the analysis and the whole rest of the cycle—for e-mail, you look at what is happening and notice: Spam has happened, oh dear! What went wrong? One of our social assumptions was wrong, namely, that everybody is friendly and will only send e-mail to

---

[9] FTP is File Transfer Protocol, for transferring files over the Internet (Postel and Reynolds 1985), and Gopher is a pre-Web Internet protocol that was menu-based rather than hypertext-based (Anklesaria et al. 1993).

[10] Berners-Lee is referring to the transformation of URIs as "Universal Resource Identifiers" (Berners-Lee 1994) to "Uniform Resource Identifiers" (Berners-Lee, Fielding, and Masinter 2005) in the final IETF Internet standard.

[11] The "Web Science Cycle" is illustrated by Berners-Lee in his 2007 presentation "Web Science: The Process of Designing Things in a Very Large Space." Available at http://www.w3.org/2007/Talks/0509-www-keynote-tbl/.

[12] SMTP is the Simple Mail Transfer Protocol, the heavily deployed IETF standard for delivering e-mail (Postel 1982).

[13] IMAP is the Internet Message Access Protocol, used by client software to modify (delete, send) e-mail stored on a server (Crispin 1994).

another person when the other person wants to read it. So the academic assumption is broken, and we have to redesign e-mail. Interestingly, no one has really succeeded in redesigning either the social or the technical piece of mail to make spam go away. So there's an example: there's a design piece and an analysis piece, an engineering piece and a science piece, with one being done on the microscopic system and the other being done on the macroscopic system, and we're missing a lot of the mathematics that would let us understand the connection between the two levels.

Harry Halpin: What is the role of philosophy in Web Science? Is there such a thing as a philosophy of the Web?

Tim Berners-Lee: An awful lot of philosophy in the past has been wasted, as it was done before we understood evolution. We were trying to understand emotions, and now we can point to evolution producing mammals with emotions. So a lot of philosophy in the past is inapplicable. A lot of people might say that philosophy is irrelevant to daily life, but if a W3C Working Group[14] stops and people start arguing about what things really mean—people refuse to play the game, refuse to say what terms mean, and they don't do their job to define a protocol properly—then it's a philosophical task to point out to them that this is important. Also, philosophy may be necessary to explain what happens when the legal system hits the Web. When you make a web-page you can link to anything, you can write anything about it. But when a lawyer comes along and reserves the right to charge you to link to their page, then in a way it's a philosophical question, as you have to tie linking to the way the protocol is defined over a name as just a reference, something that has never been controlled over the millennia.[15] Systems where you control names haven't worked so far, and so you need the philosophy to show how these protocols are grounded out in history and in concepts for using names that lawyers understand.

Alexandre Monnin: What do you expect from the philosophy of the Web?

Tim Berners-Lee: What I would like for philosophers to do is to work diligently and to produce very nice documents that describe to people like computer scientists how things work in a simple way. What happens when you click on a link? Quite a lot of that is philosophy. So, I'd like for you to have enough of a body of understanding that when people in a Working

---

[14] A Working Group is a group composed of a group of individuals that create Internet and Web standards at the IETF and W3C, respectively.

[15] There has been extensive discussion of the use of URIs as a means of reference versus a means of accessing web pages between engineers and artificial intelligence researchers (Hayes and Halpin 2008).

Group stop and say, "Wait, this doesn't match what I learned from Wittgenstein" then you can say, "No, please go read this pamphlet, it's about philosophical engineering and it explains the philosophy of what you're doing, so you won't find Wittgenstein very useful in this case or these are the bits that you will find useful." So if you can produce enough discussion and understanding so that we don't have to stop work for philosophical discussions and we can rely on philosophy being there, that would be excellent.

References

Anklesaria, Farhad, Mark McCahill, Paul Linder, Daniel Johnson, Daniel Torrey, and Bob Alberti. 1993. "IETF RFC 1436—The Internet Gopher Protocol (a Distributed Document Search and Retrieval Protocol)." http://www.ietf.org/rfc/rfc1436.txt.

Berners-Lee, Tim. 1994. "IETF RFC 1630—Universal Resource Identifiers in WWW. A Unifying Syntax for the Expression of Names and Addresses of Objects on the Network as Used in the World-Wide Web (URI)." http://www.ietf.org/rfc/rfc1630.txt.

———. 2003. "Re: New Issue—Meaning of URIs in RDF Documents." Message to www-tag@w3.org Mailing List. http://lists.w3.org/Archives/Public/www-tag/2003Jul/0158.html.

Berners-Lee, Tim, Roy Fielding, and Henrik Frystyk. 1996. "IETF RFC 1945—Hypertext Transfer Protocol—HTTP/1.0." http://www.ietf.org/rfc/rfc1945.txt.

Berners-Lee, Tim, Roy Fielding, and Larry Masinter. 2005. "IETF RFC 3986—Uniform Resource Identifier (URI): Generic Syntax." http://www.ietf.org/rfc/rfc3986.txt.

Berners-Lee, Tim, Roy Fielding, and Mark McCahill. 1994. "IETF RFC 1738—Uniform Resource Locators (URL)." http://www.ietf.org/rfc/rfc1738.txt.

Berners-Lee, Tim, Wendy Hall, James Hendler, Nigel Shadbolt, and Danny Weitzner. 2006. "Creating a Science of the Web." *Science* 313, no. 5788:769–71.

Crispin, Mark. 1994. "IETF RFC 1730—Internet Message Access Protocol." http://www.ietf.org/rfc/rfc1730.txt.

Hayes, Patrick, and Harry Halpin. 2008. "In Defense of Ambiguity." *International Journal of Semantic Web and Information Systems* 4, no. 3:1–18.

Postel, Jon. 1982. "IETF RFC 821—Simple Mail Transfer Protocol." http://www.ietf.org/rfc/rfc821.txt.

Postel, Jon, and Joyce Reynolds. 1985. "IETF RFC 959—File Transfer Protocol: FTP." http://www.ietf.org/rfc/rfc959.txt.

Wilks, Yorick. 2008. "The Semantic Web: Apotheosis of Annotation, but What Are Its Semantics?" *IEEE Intelligent Systems* 23, no. 3:41–9.

CHAPTER 13

AFTERWORD: WEB PHILOSOPHY

BERNARD STIEGLER

If we want to ask what "philosophical engineering" is, we need to ask ourselves, first, what "engineering" is and, second, what relationship philosophy has with the elementary features of engineering. The question of engineering is not exactly the same question raised by Archimedes, even though it is quite often said that Archimedes was the first engineer. Archimedes himself was the first example of a technician who anticipates the appearance of engineering by enacting a *mechanē*, a *mechanē* that is grounded on an *epistēmē*, an empirical and practical mechanics built from mathematics and physics. Archimedes, however, prefigures the constitution of engineering taken as the ending of an ontological opposition—that is, the very basis of ontology itself—which posits that the primary principle of rational knowledge, and thus science, is that it constitutes itself by that which is necessary ("that which cannot be other than it is") and that attains this certain *apodictic*[1] view purified of any technicity.

Philosophy constitutes itself not only by differentiating but even by opposing *epistēmē* and *technē*, although originally those two notions were formerly synonymous, as Heidegger reminded us, and as can be found in Homer. *Epistēmē*, in the philosophical sense of Socrates and those whose followed him, is grounded on *apodicticity* that in turn presupposes an idealization, which believes itself or asserts itself to be—as Immanuel Kant would formulate much later—a priori and thus transcendental: not coming from experience and thus having nothing to do with the order of existence,

---

[1] This Aristotelian term means being logically demonstrated. In this chapter, the footnotes are by the volume editors, Harry Halpin and Alexandre Monnin. The translation, from the French, is by Louis Morrell and Harry Halpin.

Philosophical Engineering: Toward a Philosophy of the Web, First Edition. Edited by Harry Halpin and Alexandre Monnin. Chapters © 2014 The Authors except for Chapters 1, 2, 3, 12, and 13 (all © 2014 John Wiley & Sons, Ltd.). Book compilation © 2014 Blackwell Publishing Ltd and Metaphilosophy LLC. Published 2014 by Blackwell Publishing Ltd.

but concerning rather the order of consistence, objects which do not exist but which have consistence, and these have been called since Plato "the ideals" (Kant 1929).

There cannot be knowledge without ideals; this is what Plato asserts and I myself continue to assert. However, I add that ideality today must be thought not only via Plato but also via Freud, and at an inevitable moment against Plato, against his opposition of the soul and the body and not only of technics and knowledge—which is also to say of becoming and being—an opposition that constitutes an ontology that I would deny, as I explain later.

The question of idealization is that of desire, insofar as it is itself a dynamic process. This question of desire as dynamic is reintroduced by Aristotle—reintroduced insofar as it springs into being with Socrates in the *Symposium*, where Diotima paints knowledge as a child of Eros (Plato 1993). Spinoza develops and explores in a systematic fashion this Aristotelian inspiration, which is at that point an *anamnesis* of Socrates.[2] Freud and his successors—Winnicott, Lacan, Marcuse, and a number of others who incidentally do not necessarily hail from psychoanalysis; Marcuse, for instance, is a philosopher—bring a new turn to these questions of desire and idealization, by introducing concepts such as sublimation (which becomes a fundamental element in understanding *libido sciendi*),[3] the unconscious (which roots the soul within the body), fantasy, and so on. I consider these theories to be considerably unfinished, incomplete, and therefore unsatisfactory, but they still constitute contributions to a return to thinking through the possible conditions of all knowledge.

Be that as it may, to think through philosophical engineering is not only to think through the relationship of philosophy to technology but to think through technology in a sense where the engineer is understood as a figure of knowledge coming to the fore not so much in the Renaissance as with the French and Prussian engineering schools in the late nineteenth century, through which knowledge comes to be one of the main factors of economic competition, and so thereby escaping the grip of Christian clerics, the representatives of the Symbolic, which itself had emerged out of the ontotheology of Christianity. Contrary to the Platonic point of view, I share with the late Husserl a thesis that phenomenology has never understood: the exteriorization of knowledge is the condition of the constitution of knowledge (Husserl 1970).

---

[2] *Anamnesis* can be thought of in its Platonic sense as knowledge thought of as remembrance, of making present to oneself that which is necessary.

[3] Roughly, the desire for knowledge.

This exteriorization of skillful knowledge means that full-fledged knowledge[4] is not identical to cognition—if by cognition one means a behavior adaptive to an environment that is understood as a defining element of living beings in general with a view to including even artifactual automatism. Skillful knowledge, even when it does not come to constitute full-fledged knowledge, is not simply mere cognition.

Knowing how to act, knowing how to live, and knowing how to think rationally[5] are not mere forms of cognition, because they are grounded on an exteriorization process that cannot be supervened on lower-level cognition, such as animal cognition, for instance—although, obviously, it is the great apes that put into motion the exteriorization process before hominization even began. Let us not return to the question of whether man is or isn't an animal; the question is to determine which processes and the dynamic factors reside within what Simondon calls the individuation processes, and how a transition from biological individuation to social individuation can be achieved (Simondon 1989). As this latter psychological and social individuation is also technological, clearly this is the point where the question of a philosophical engineering is to be asked, since such a point of view leads to us posit that all psychological individuation is always linked with a technical or technological individuation. Cognitivist byzantinism will not make any push forward in such a field: cognitivism is precisely what for whole decades has hampered the exploration of the role of artifactual exteriorization within the constitution of knowledge.

Drawing from Husserl and Derrida, but also putting into question a certain number of Derrida's presuppositions (such as his assertion that the *trace* needs to be approached from a quasi-transcendental standpoint, a presupposition that must be avoided like the plague, since it limits the question to that of "the" trace when there are only traces and it is this empirical multiplicity that needs to be thought out), one can differentiate tertiary retentions from primary and secondary retentions, and therefore posit that there is no *arche-trace* without the interplay of plural traces (Derrida 1976).[6] It is this multiplicity of traces, which includes cerebral-mnemonic traces, that is spawned from the tertiary retentions produced by exteriorization, which itself also conditions the individuation of the cerebral apparatus, as well as the psychological apparatus and the social

[4] In this translation, the French term *connaissance* is rendered as "full-fledged knowledge" in this paragraph, while *savoir* is rendered as "skillful knowledge" in this paragraph. In the rest of the chapter, *savoir* is rendered simply as "knowledge," since it is no longer contrasted with *connaissance*.

[5] The original French has this form as a triad: "*le savoir-faire, le savoir-vivre et le savoir penser rationnellement.*"

[6] The *arche-trace* is the "originary trace" (Derrida 1976, 61).

apparatus, something which belongs to what I call a "general organology" (Stiegler 2004).[7]

As an interplay of three different types of organs—physio-logical, techno-logical, and organizational, which is to say socio-logical[8]—exteriorization is that which is produced through circuits of traces between all of these. It is circuits of exteriorization and interiorization that produce the three kinds of retentions and condition the effective reality of the transindividuation processes that are characteristic of the process of collective individuation. It is those transindividuation processes that explain how from the absolutely singular experience wherein "I think by myself," that which a phenomenologist would call a first-person experience, I am able to produce common knowledge, like that of the infinite "we" of geometry, the "we" of the community of geometers, which is an infinite community: a process of infinite collective individuation, forming a community that will never be closed, as Husserl would say, something that is only possible within the tertiarization (exteriorization) of secondary and primary retentions (Husserl 1970). Although further exploration of this is beyond our present scope, such a standpoint requires us to revisit Kant's transcendental deduction and add a fourth synthesis, one that would presuppose a new conception of schema and schematism as well (Kant 1929). If I specify such a point here, it is because the questions raised by the World Wide Web Consortium[9] about the categorization of digital data and metadata are a matter of schema in the Kantian sense—that is, of a projection between a concept of understanding and the data of intuition ("intuition" being used as well in the Kantian sense of being constituted by—and in—space and time).

Intuition as well as understanding are "virtualized" and so in this sense "exteriorized," and that schematism is taken up by a process of automation, since the goal of such an exteriorization is precisely to render possible the performance of mechanical calculations by automatons. All this belongs uniquely to the contemporaneous stage of "grammatization." Before returning to grammatization, I stipulate that tertiarization is only possible through exteriorization and re-interiorization (or to say the same conversely, as interiorization and re-exteriorization), something that can be represented by a spiral. We need to go back and read Jacob von Uexküll and the question of the sensor-motor loop with such a viewpoint; we can show that von Uexküll's loops, from the moment

---

[7] "General organology" is a term used by Stiegler to describe the study of the joint individuation of the psychological life of individuals, social groups, and technology.

[8] The hyphenation is preserved from the French: "*physio-logiques, techno-logiques et organisationnels, c'est à dire socio-logiques.*"

[9] The World Wide Web Consortium (W3C) is the Web standards body founded by Tim Berners-Lee, and is known for producing data standards such as XML (eXtensible Markup Language) and the Semantic Web standards (Berners-Lee, Hendler, and Lassila 2001) for metadata.

where tertiary retentions take place, produce infinitely long circuits of transindividuation (von Uexküll 2011). It is within the social body of the collective that the individual is formed, through the stratification, depositing, and accumulation of tertiary retentions, collective secondary retentions that have been exteriorized and metastasized under the form of transindividuation circuits by generations since the dawn of humanity, from sharpened flints to the Web and beyond.

Before talking specifically of the Web and ontology in the age of digital categorization, I would like to bring to mind something that everyone knows, to wit, that the question of the trace and so the question of writing (since every tertiary retention is a writing), which is for Plato and Socrates the question of the *pharmakon,* builds upon the general organology mentioned earlier. So, concerning writing, the question is neither to exclude it from thought, as Derrida wrongly claims that Plato did, nor to contain it in the shackles of dialectics, as Socrates offers to do in *Phaedrus* (Derrida 1981). Such a dialectics, which would consist of a therapeutics that would make use of writing only by controlling it through and through, actually presupposes writing, so that writing *always already* contaminates dialectics, as Derrida would say, and therefore cannot be controlled. What Derrida may have said from a very speculative standpoint, we today know from a much more empirical and scientific, anthropological, archaeological, and computational standpoint, and from the standpoint of what neuroscience can produce best as well.

Be that as it may, the literal tertiary retention that emerges with the appearance of the letter, through the process of "grammatization" as Auroux conceives of it (Auroux 1994), goes on today through what Auroux calls "the industries of language," and what Frédéric Kaplan describes as "linguistic capitalism," of which Google is the most typical representative—but there are many others (Kaplan 2011). The question raised by this grammatization process today is that of a philosophical engineering, something that conditions the constitution of knowledge in a general manner, not only philosophical dialectics, but all forms of knowledge, including knowing how to live[10] and knowing how to act.[11]

The thinking that this question provokes takes place in a time that is characterized by two fundamental traits. First, memory has become an economic and industrial activity, a source of profit. Second, this economic activity is undertaken through this whole new form of writing that is digital technology. Digital technology is the most current form of writing—nothing more, nothing less. Yet it is a form of writing that belongs to the order of what Marx, in the *Grundrisse,* describes as automation, a new process where science comes to be in the service of economics, of the amortization of capital's investment, in a direction that leads to the

---

[10] In French, *savoir-vivre.*
[11] In French, *savoir-faire.*

destruction of all knowledge, which Marx at the time called proletariza-tion (Marx 1973).

To this, we need to add that if we turn now to the question of the Web itself—and not of digital technology in general—the Web is a nexus of thousands of questions today, in particular that of the relation between the Semantic Web[12] and the Social Web,[13] and therefore of the relation between language, algorithm, and society (Berners-Lee, Hendler, and Lassila 2001). Then if we turn to this question, we see that the Social Web is what cannot be reduced to calculation, what is not soluble by the algo-rithmic, a position I uphold without opposing the social and algorithmic web. The Social Web is fundamentally communities of annotators using algorithms in order to verify hypotheses, the extraction of data through visualization and techniques of analysis such as signal processing, and all things of this nature, with all of these activities concretizing as inscriptions and annotations from readers within a database of annotations that has created a new heuristic and hermeneutic space.[14] This question of articula-tion between what are now called the Social Web and the Semantic Web is also the question of automatization and the question of the autonomy of the interpreter—interpreters are only such if they interpret in an auton-omous manner, that is, by giving themselves their own rules of reading and thereby creating a new reading whenever they succeed in producing a conclusive transindividuation, for instance by creating a school of liter-ary criticism, or by becoming writers themselves. This is for instance the subject matter of Henry James's *Figure in the Carpet* (James 1916).[15]

Today, then, the question concerning Web engineering is to determine how to deal with the relation between autonomy and automatization in a context that is highly pharmacological.[16] More specifically, the question is to know how to deal with this relation by revisiting the question of ontology as it is framed in philosophy: the problem of universals from Aristotle to Heidegger, through Husserl, and in so many other dimensions of course, including Kant—and by incorporating the fact that the indi-viduation of the retention system is what conditions the constitution of ideals. This retention system is not simply an external condition to this

[12] The Semantic Web has been described as a web where machine-readable data are given meaning (Berners-Lee, Hendler, and Lassila 2001). It is currently a standardization effort in the W3C.

[13] The Social Web is often thought of as Facebook and other technologies that directly support and work via social relationships.

[14] Stiegler has held this position since he began work at the Bibliothèque Nationale de France with Jacques Virbel and Philippe Aigrain in 1989, where it was phrased as "learned communities of readers" rather than as communities of annotators.

[15] Stiegler may touch on this theme in greater detail soon in a book that is tentatively entitled *Mystagogies*.

[16] By "pharmacological" Stiegler refers to the ancient Greek concept of *pharmakon*, describing an open-ended process of change mediated via technical objects, a process that may become either toxic or a cure for the problems of our current condition.

constitution but a process of psychological and collective individuation that is grounded on technical individuation, within which technics are not only an environment but constitute the condition of projection of desire of individuals from a given community, since behind all of this the question of the idea as ideal, as desire, is raised.

Ontology aims at describing "what is," which means: what is truth, either from a fundamental standpoint, through categories that would then be the a priori conditions for any kind of predication, or as a regional ontology that would describe fields of the empirical within the real that conform with fundamental ontology but also dovetail with materialist ontologies. Whatever the kind, ontology tends to posit that there is some being (to use Heideggerian language) and that these beings have an essence, which itself is ideal and invariable in the sense that the ideal constitutes what Husserl calls the *eidetic core*, which is the intentional beginning of a phenomenon that any subject to which a phenomenon appears must take as identical to itself (Husserl 1982). Reading Aristotle, then Husserl, turning to Simondon and Nietzsche: this manner of tackling these issues is as a whole insufficient nowadays. This does not mean that it is inadequate but that it needs to be entirely reframed—and I think that the word "ontology" has to be discarded; I think it would be better to adopt "idealization," just as a geographer idealizes a landscape by extracting its *eidē*, by producing an idea from it, an aspect, *eidos*.

Here we must discuss the question of the neurological condition of thinking about numbers and mathematics more generally. By referring to the works of Changeux and Berthoz, Dehaene claims to lay out the features of a mathematical bio-neuro-ontology or a physical bio-neuro-ontology (Dehaene 2009). We could then say that if Meno, by considering the tracings that Socrates forms in the sand, can ultimately arrive at the discovery of the conditions for calculating a square surface, it is very simply because his brain already contains this "in-formation," this coming into form of his neurons through his biological heritage, that is to say, through environmental adaptation (Plato 1976).

I think that such a way of thinking is simultaneously very interesting, indubitably supported by documentations and laboratory observations that are fully worth the attention, and at the same time wholly insufficient and so, to phrase it in a slightly brutal manner, perfectly naive. To understand this, we need to take a detour through the discussion that takes place, in a rather oblique and even almost occult manner, between Stanislas Dehaene and Maryanne Wolf. Wolf, in her reflections on reading, says that we were never born to read (Wolf 2007). In saying this, she radicalizes a position that is adumbrated but never carried out as such by Deheane, to wit, that the ability to read presupposes a reorganization of the cerebral circuits, of cortical and subcortical organization, which consists in destroying certain former aspects of that organization in order to build new ones (Dehaene et al. 2005). Wolf positively tells us that before the acquisition of writing, there are conditions for acquisition which are neural in nature, and which are already existing organizations of the brain (Wolf 2007). Yet the real question is that of their destruction and replacement by new organization—a kind of cerebral rearrangement that happens in accordance with the organologic[17] specificities of a new technique, in this case, writing—and the acquisition of writing; this reorganization of the brain through writing is what enables a new way of thinking, feeling, memorizing, and such to appear.

In other words, the exteriority of artificial organs is a dynamic process that is not the result of the brain's activity or of its projection to an outside but is actually what coproduces the becoming of the brain, what someone like Simondon would have called the individuation of the brain or the ontogenesis of the brain (Simondon 1989). Maryanne Wolf is not only a neuroscientist but also a mother, who observes all of these phenomena in her own children not only from an analytical standpoint but also from the standpoint of a mother who asks herself how the digital brain will be able to withstand, to grow, without producing negative effects. If bodies like the World Wide Web Consortium do not take on this kind of question, these organizations cannot reach very far. It is on Dehaene that Wolf draws, as Dehaene raises the question of reading and the neurological abilities needed to explain reading (Dehaene et al. 2005). Yet Dehaene does not do as Wolf does, since he raises the question by stating that today the brain is forced to interiorize reading and a number of circuits that were not inscribed into the brain from the outset, because the brain did not have enough time to produce such circuits itself (Dehaene 2009). For me, this is an unfathomable stupidity, a result that would prefer to make everything reducible to the regional ontology of the neurosciences.

Deeper even than ontologies, which are only the surface effects of transindividuation processes that have metastasized and that in so doing have erased the movement that produced them, there are individuation processes. Those individuation processes spawn metastasis processes that are idealizations of the present state of relations, driven toward upholding the system in its greatest potentiality, that is to say, as its negentropic[18] maximum, loaded with negentropy in the sense in which Simondon speaks of a pre-individual loaded with potential (Simondon 1989).

This is what philosophical engineering is in the era of the Web.

To conclude, I would say that behind all of these questions, underlying what is at stake, is how to think about the relations between psychological, technical, and collective individuations in a context of engineering—where

---

[17] A reference to Stiegler's theory of a "general organology" (Stiegler 2004).

[18] Negentropy is the ability of a system, such as a living system, to increase its organization relative to the entropy around it, and so to counter the immediate effects of the tendency toward entropy as given by the Second Law of Thermodynamics (Schrödinger 1944).

engineers are in the service of the economic activity of linguistic capitalism, which is no longer merely linguistic but has become the digital capitalism of which Google is only a dimension—and which is towered over by the question of automatization.

This question of automatization is that of the generalization of the algorithmic, of which informatics is an incarnation. A Vaucanson automaton is already algorithmic, it is a physical materialization that eventually allows for the repetition of a process.[19] Because automatism appears at the origin of industrial society and of industrial mechanization, and today develops not only machines but also devices that can become microscopic (even nanometric), the issue becomes framed in entirely unprecedented dimensions. For instance, technological automatism can become the reason behind psychological automatism and neurological automatism, and even coalesce with them—these are the true stakes of what is called the NBIC convergence.[20]

In such an age, the absolutely unprecedented question that is raised is that of the relation between automatization and autonomy. Behind this, Enlightenment philosophy itself is at stake. What did the Enlightenment philosophers want? They wanted to extend autonomy, and this extension of autonomy was grounded on the development of the "Republic of Letters."[21] This republic itself is what allowed for the constitution of what Kant called "the reading public," the *res publica* (Kant 1784). Yet the Enlightenment never gave thought to the letter itself, even from the standpoint of social mathematics, although Condorcet and some others did circle around such issues (Condorcet 1795). I have tried to show in *Le temps du cinéma* that Kant's transcendental deduction renders it impossible to think through the Republic of Letters and more generally the tertiary retention (Stiegler 2010). By contrast, we know then that the question, not only of the alphabetic letter but also of current technology, is intractable, and so the question cannot dissolve within pure reason, either theoretical or practical. Reason is impure, and this impure reason is for the most part a computational reason that is nevertheless not reducible to computation itself. This reason is a reason of desire, because reason means motive and desire. It is a reason of shadows as well as of enlightenment, and its light and shadows are also numbers and shadows[22] in the age of

---

[19] Jacques de Vaucanson was a French inventor in the eighteenth century who created many innovative mechanical automatons that often mimicked biological forms, such as ducks.

[20] NBIC is an abbreviation for "nanotechnology, biotechnology, information technology, and cognitive science."

[21] This is a reference to the community of scholars, usually unaffiliated and unfunded, that arose in the seventeenth and eighteenth centuries, whose members would offer each other help with research and frequently correspond over scientific and political matters (Goodman 1994).

[22] In the original French, the terms "shadows" and "numbers" are a play on words: *ombres* (shadows) and *nombres* (numbers).

light-speed optoelectronics, which in turn requires a new physics of philosophy—a physics of photons that cancels time and provides techno-logical automatism with an unprecedented power to connect with psycho-logical automatism.

Numbers allow us to project idealizations into all kinds of domains, from the Renaissance, in the field of mathematical physics, then later of social mathematics, and so on; and these projections, which are the basis of experimental sciences, and of modern science insofar as it is computa-tional, require processes—as stated by Husserl in *The Crisis of European Sciences* (Husserl 1982)—of the automatization of reasoning, and the problem is that these processes of reasoning automization lead to what Adorno calls the dialectics of reason: to wit, the reversal of reason (Adorno and Horkheimer 2002). The dialectics of reason lead to irrationality—something that we are now unfortunately living through, through the financial crisis and its consequences, even though it seems that few people have actually drawn the proper consequences from this crisis, on the philosophical, mathematical, or economic level, and so on.[23]

Be that as it may, the real question that is to be raised is that of a philosophy of enlightenment, of shadows and of numbers in the digital age.[24] This must include speed, in particular as regards light-speed, since all these Web technologies are based on optoelectronic technology, a precise light-time that rests on a mobility without mass, on photons that do not fit in Newtonian mechanics but fit in Broglie's wave mechanics.[25] All the questions that Bachelard opened at the dawn of the twentieth century need to be thought out on absolutely new grounds, since the artifacture that I call "general organology" can no longer be avoided, and the question is to be framed as a pharmacology where the question is, as always, how to produce autonomy from heteronomy. Not against heteronomy, as was believed by philosophers up to Kant and even beyond, but with heteronomy, a heteronomy that negatively is a pharmacological poison, precisely because it short-circuits autonomy, but that when one pushes for a therapeutic can become a positive pharmacology—that is, a future for an investment in the social, the amorous, the affective, the scholarly, and also the economic, that is, for a new industrial society.

These are the stakes of Enlightenment philosophy in the era of the Web.

---

[23] This sentence starts with a reference to *Dialectique de la raison,* the original French mistranslation of the title of Adorno and Horkheimer's *Dialectik der Aufklärung,* translated in English as *Dialectic of Enlightenment* (Adorno and Horkheimer 2002).

[24] The original French of "the Enlightenment, shadows, and numbers" is *des Lumières, des ombres, et des nombres,* which repeats the play on words from earlier.

[25] Louis de Broglie was a French physicist who introduced the idea that all matter can act as both particles and waves, as Einstein discovered in the case of photons.

References

Adorno, Theodor, and Max Horkheimer. 2002 [1944]. *Dialectic of Enlightenment*. Translated by Edmund Jephcott. Stanford: Stanford University Press.

Auroux, Sylvain. 1994. *La révolution technologique de la grammatisation*. Liège: Mardaga.

Berners-Lee, Tim, James Hendler, and Ora Lassila. 2001. "The Semantic Web." *Scientific American* 5:29–37.

Condorcet, Marie-Jean-Antoine-Nicolas Caritat, Marquis de. 1795. *Outlines of an Historical View of the Progress of the Human Mind*. Philadelphia: Online Library of Liberty. Available at http://oll .libertyfund.org/index.php?option=com_staticxt&staticfile=show .php%3Ftitle=1669.

Dehaene, Stanislas. 2009. *Reading in the Brain*. New York: Penguin.

Dehaene, Stanislas, Laurent Cohen, Mariano Sigman, and Fabien Vinckier. 2005. "The Neural Code for Written Words: A Proposal." *Trends in Cognitive Sciences* 9, no. 7:335–41.

Derrida, Jacques. 1976. *Of Grammatology*. Translated by Gayatri Chakravorty Spivak. London: Johns Hopkins University Press.

———. 1981. "Plato's Pharmacy." In *Dissemination*, 61–172. Translated by Barbara Johnson. London: Athlone Press.

Goodman, Dena. 1994. *The Republic of Letters: A Cultural History of the French Enlightenment*. Ithaca: Cornell University Press.

Husserl, Edmund. 1970 [1936, 1954]. *The Crisis of European Sciences and Transcendental Philosophy*. Translated by David Carr. Evanston, Ill.: Northwestern University Press.

———. 1982 [1913]. *Ideas Pertaining to a Pure Phenomenology and to a Phenomenological Philosophy: General Introduction to a Pure Phenomenology*. Translated by Fred Kersten. The Hague: Nijhoff.

James, Henry. 1916. *The Figure in the Carpet*. London: Martin Secker.

Kant, Immanuel. 1784. "What Is Enlightenment?" Translated by Mary C. Smith. Available at http://www.columbia.edu/acis/ets/CCREAD/ etscc/kant.html.

———. 1929. *The Critique of Pure Reason*. Translated by Norman Kemp Smith. London: Macmillan.

Kaplan, Frédéric. 2011. *Google et le capitalisme linguistique*. Available at https://fkaplan.wordpress.com/2011/09/07/google-et-le-capitalisme -linguistique/.

Marx, Karl. 1973. *Grundrisse: Foundations of the Critique of Political Economy*. Translated by Martin Nicolaus. Harmondsworth: Penguin.

Plato. 1976. *Meno*. Translated by G. M. A. Strube. Indianapolis: Hackett.

———. 1993. *The Symposium*. Translated by R. E. Allen. New Haven: Yale University Press.

Schrödinger, Erwin. 1944. *What Is Life?* Cambridge: Cambridge University Press.

Simondon, Gilbert. 1989. *L'individuation psychique et collective.* Paris: Aubier.

Stiegler, Bernard. 2004. *De la misère symbolique, tome 1: L'époque hyperindustrielle.* Paris: Galilée.

———. 2010. *Technics and Time, 3: Cinematic Time and the Question of Malaise.* Translated by Stephen Barker. Palo Alto, Calif.: Stanford University Press.

von Uexküll, Jacob. 2011 [1934]. *A Foray into the Worlds of Animals and Humans.* Translated by Joseph D. O'Neil. Minneapolis: University of Minnesota Press.

Wolf, Maryanne. 2007. *Proust and the Squid: The Story and Science of the Reading Brain.* New York: HarperCollins.

INDEX

Philosophical Engineering: Toward a Philosophy of the Web, First Edition. Edited by
Harry Halpin and Alexandre Monnin. Chapters © 2014 The Authors except for Chapters
1, 2, 3, 12, and 13 (all © 2014 John Wiley & Sons, Ltd.). Book compilation © 2014 Blackwell
Publishing Ltd and Metaphilosophy LLC. Published 2014 by Blackwell Publishing Ltd.